Henry Beaumont Small

The animals of North America

Henry Beaumont Small

The animals of North America

ISBN/EAN: 9783337228538

Printed in Europe, USA, Canada, Australia, Japan

Cover: Foto ©berggeist007 / pixelio.de

More available books at **www.hansebooks.com**

THE

ANIMALS OF NORTH AMERICA.

"Non injussa cano   *   *   *   *
Cetera, quæ vacuas tenuissent carmine mentes
Omnia jam vulgata."

BY

H. BEAUMONT SMALL, S,C.L.

Montreal:
PRINTED BY JOHN LOVELL, ST. NICHOLAS STREET.
1864.

CLASSIFICATION ADOPTED.

Bimana,.......................*Two handed.*
Quadrumana,..................*Four handed.*
Cheiroptera,..................*Finger-winged.*
Insectivora,..................*Insect-devouring.*
Carnivora,....................*Flesh-devouring.*
Rodentia,.....................*Gnawing.*
Marsupialia,..................*Pouched.*
Ruminantia,...................*Cud-chewing.*
Pachydermata,.................*Thick-skinned.*
Cetacea,......................*Whales.*

ORDER.   GENUS.   SPECIES.

AUTHOR'S PREFACE.

THERE are two reasons which have induced the Author to publish the following pages. The first is, that as the pleasing study of Natural History ought to be extensively introduced into institutions of learning, yet the generality of books already in circulation on this subject, present to the mind of the student either too great an amount of detail, or else include in a single volume, necessarily meagre, the whole Animal Kingdom. The second is, that a growing desire for further acquaintance with this study is felt among a large and increasing class of intelligent readers, who have not the facilities for using books of reference which *savants* have. In a country like this where a man is brought into contact with mere nature, teeming with unsuspected wealth, of what incalculable advantage is it to have, if it be but the rudiments of a science which will tell him the properties, and therefore the value of its animals and natural productions. He whose mind is relaxed and wearied, after the hours of business, will not sit dreaming over impossible scenes of pleasure, or go for amusement to haunts of coarse excitement, if his interest is once awakened in some study fitted to keep the mind in health. To gratify this desire to some extent, and to assist students in this department, is the object of the present work. Much of the matter is original, the result of a long and somewhat extensive familiarity with the science. Much also has been gathered from reliable sources; the whole divested as much as possible of all asperities, in the form of scientific names, which so often deter beginners.

In conclusion, the author takes the opportunity of expressing his thanks to Sir Wm. Logan, Mr. Billings, and the Natural History Society of Montreal, the use of whose libraries was kindly tendered and accepted, and to all those who have evinced an interest in the progress of the work. Should the success of this volume on the Mammalia warrant the experiment, others will follow in due course, comprising the remainder of the system.

H. BEAUMONT SMALL.

MONTREAL, September 1, 1864.

THE

ANIMALS OF NORTH AMERICA.

CHAPTER I.

LIMITATION OF SPECIES—CENTRES OF CREATION—FACTS AND FICTION—FAUNA OF AMERICA—OBJECT OF THIS WORK—BATS—THEIR HABITS—THE SHREW-MOLE—ANECDOTE—THE STAR-NOSED MOLE—THE SHREW-MOUSE.

One of the most remarkable things that strikes even a casual observer, in taking a view of the Animal kingdom, is the manner in which species are distributed over the globe; but to understand this, it is necessary to look at the different influences which circumstances exercise over them. Each division of the world has a *fauna* (or group of animals) peculiar to itself, characterized by some remarkable species found there only. This has been termed the "limitation or colonization" of species, and has given rise to many theories;—one affirming that each race originated in the spot destined for it; another, that the same country saw the birth of every distinct race, which, migrating, and leaving no trace of their passage, colonized as it were, eastward and westward, and in the island groups of the Southern Ocean, as either place was best adapted for their development; while some again maintain that there was originally but one form created, from which all others have risen *ad infinitum*, being so changed by climate and circumstances, as to eventually cause distinct species, generating fresh ones in their turn, and terminating with the human family as the masterpiece of this successive formation.

The most natural supposition is, that the all-wise Creator placed each species where it was permanently destined to live; and that from these different "centres of creation," combinations have so multiplied between contiguous regions, as to form the various races of animal life. When we find a country possessing a group or groups of animals not found elsewhere, we may at once set down that as being the *centre of a peculiar* creation. In the location of many species, nature has placed various limits, and the spaces occupied by them are most unequal. For example: the Kangaroo and Ornithorhyncus are confined to New Holland; the Grizzly Bear to the Rocky Mountains; the Dodo, now extinct, to the Mauritius; whilst the Swallow, the Crow, and the Fox, extend to every known region. The principal cause of "limitation" is doubtless connected with the unequal temperature of localities; certain species which thrive in one climate, perishing under the influence of another;—also the nature of vegetation in one country, and the absence of it in another,—as in the Polar regions,—confining to the former the larger beasts of prey, dependent on herbivorous animals as their food, with the exception of, in the latter, those that subsist on fish. The number of species increase as we near the tropics, and there it is where Nature has been most lavish in the diversity of life, beauty of color, strangeness of form, and greatness of proportion. The present total number of living species which has been satisfactorily made out and ascertained, exceeds, according to Agassiz, 50,000!

If the time ever comes when the facts of Natural History are given without the admixture of fable, then this branch of science will be more readily advanced in improvement than can be readily hoped for, so long as imagination is allowed to take the place of actual observation. Modern writers continue to intermingle so much of what is barely possible with the little attested, as to give an air of doubt to the whole. We are nearer the truth when we admit our ignorance, than when we embrace an erroneous hypothesis; for we have but to learn in the first case when the truth is developed; while in the latter, we have to unlearn before we can learn. This experience always proves to be the greatest difficulty to a learner. Many of the narratives of the older naturalists are little more than amusing fables. To deduce the leading characteristics of an animal from a minute investigation of its physical construction, to watch its habits in its native haunts, formed no part of the care of those who compiled books on natural history a century ago. Whatever was imperfectly known was immediately made the subject of some tale of wonder.

Some writers, unable to ascertain for themselves, accept and publish to the world the information given by trappers and travellers, in which case many errors may have arisen from the ignorance of the observer; though in addition to these errors of ignorance, there must be added a worse evil—viz: the love of the marvellous, which has contributed largely to false accounts. Godman, the well-known American Naturalist, recites an instance of this, where a trader, having given a most fictitious account of the habits of the beaver to an ardent enquirer, who carefully noted all down, remarked on the departure of the latter, that, being so annoyed by a constant enquirer, he had chosen to get rid of him by this method, viz: appearing to tell him all he knew! Such errors as this are great drawbacks to accurate students, and delude the minds of learners. The injury which the mind receives from this source is scarcely appreciable, and the false notions we form concerning the plans of Nature, are not easily afterwards eradicated.

According to Buffon, the *fauna* of America is characterized by inferiority in size when compared with that of the old world; on the other hand, it is the richest in species, none having yet been extirpated, possessing 557 mammalia, of which 480 are its own. One curious feature is, that no country has contributed so little to the stock of domestic animals, having furnished, with the exception of the llama and the turkey, no animal serviceable to man. In connection with this, however, we must remark, that a commonplace observer would be apt to imagine that the vast herds of wild cattle and horses which roam in thousands over the savannahs of Mexico and the extreme Southern States, are indigenous; little thinking that they are the descendants of the few animals the Spanish conquerors permitted to run wild, which have resumed the originality of their species.

The object of this work is to enumerate the different species of animals of the Northern Continent of America, arranged as nearly as possible according to Cuvier's system, with the introduction of certain incidents and peculiarities really authorized and reliable, and which are in many instances unknown to the majority of readers,—peculiarities which open new fields of enquiry, and lead the observer to perceive that what appears accidental in the habits of the Animal World, is the result of some unerring instinct, or some singular exercise of the perceptive powers, affording the most striking objects of contemplation to a philosophic mind.

Passing over the first family (*bimana, two-handed*) Man,—and the second (*quadrumana, four-handed*) or Monkeys, as wanting in North America, we commence with the third,—

CHEIROPTERA, (*wing-handed*). The Bat.—(*Vespertilio.*)

Description.—Ears broad; the anterior and posterior extremities connected by a more or less naked expansion of the skin, or a membrane including the tail, adapted for the purpose of flight: prey upon the wing: nocturnal in their habits.

Few if any of the individuals of the Animal Kingdom are so singularly and curiously formed as the bat. It is described by an eminent writer as " holding a very equivocal rank in creation, and though having a marked resemblance to a quadruped, a great part of his life is spent in the air like a bird." Instead of being oviparous or egg-laying, this is a lactescent, or milk-giving animal; instead of living on grain, its food is flesh; and instead of being like a bird, a biped or two-legged animal, it is a quadruped in the true sense of the term.

Great ignorance prevailed among the ancients respecting bats. Aristotle describes them as "birds with skinny wings!" Pliny asserts that they are "birds which produce their young alive, and suckle them;" while Aldrovandus, who always has something exquisitely graphic, places them in the same family as the Ostrich, giving as his reason, that "these two species partake equally of the nature of quadrupeds!!" How, why, or from what similitude, he leaves an open subject.

The wings of the bat are formed by the extension of a fine membrane over the elongated fingers of the fore-legs, reaching as far as, and fastened to, or rather stretched over the hind-legs. As however the four fingers are involved in the membrane which forms the wings, only a little hook, called the thumb-nail, is left free. With this the animal suspends itself on any rough or uneven surface where it happens to alight; while the hind feet are also provided with claws, by which it hangs head downwards on the sides of chimneys, hollow trees, and roofs of caverns, a favorite resort, still and silent, sleeping, or perhaps nursing its young by day, till the approach of evening, when it begins its excursions in search of food.

Having neither the disposition nor the power to exercise themselves by day, bats are strictly nocturnal animals, commencing their search after insects soon after the swallow has quitted his operations for the day. Its motions, as it flits about in the dim twilight, seldom moving more than a few yards in a straight line, darting up or down, this way or that, instead of being for its mere pleasure, as many would suppose, are really its only means of procuring its living, since at every turn it seizes, or attempts to seize, some one of the insect tribe, which swarm under cover of darkness in the air. While on the wing it continually utters a low shrill cry, not unlike the squeaking of a mouse.

Naturalists have long since discovered by experiments, that bats deprived of sight, still avoided obstacles as perfectly as those with their sight entire, flying through small apertures only just large enough to admit them without touching; numerous small threads also were drawn across the room where the experiment was made at different angles, and still the blind bat would fly about in every possible direction without ever touching them. The vibration of the air striking against the impediment, was supposed to return a sound by which the animal was warned of its direction. But it has since been found that the destruction of hearing as well, made no difference in the fact, and the only theory that has been proposed to account for this curious circumstance is, that some peculiar sense is lodged in the expanded nerves of the nose.

No authentic records have ever come before the writer's notice, of the bat having been tamed; they seldom live any time in captivity, but will eat fearlessly and voraciously of raw meat; they invariably refuse the house-fly. There are a number of American species, all agreeing very nearly in habits and form, amongst which the following are mostly met with.

V. NOVEBORACENSIS (New York Bat). This species is common throughout the Northern part of the United States, and not uncommon in Canada, its range extending between the thirty-third and forty-second parallels of latitude.

V. PRUINOSUS (Hoary Bat), of a grayish color, its hair being black, tipped with white, hence its name hoary. Not common, and but little known of its habits; its range extensive, but limits not known.

V. SUBULATUS (Little Brown Bat). This species is subject to great variation in size and color; it is found all over the continent as far as 53° North latitude.

V. NOCTIVAGANS (Silver-haired Bat), color uniform black, with a sort of collar composed of white or silver tipped hairs surrounding the neck, and ascending the ears. Its history very incomplete and range not known, but is said not to extend north of Massachusetts.

V. CAROLINENSIS (Carolina Bat), glossy chestnut color; large size; interfemoral membrane not enclosing tip of the tail; range said to be from Georgia to Connecticut.

INSECTIVORA (INSECT-EATERS) is the next order, comprising only the shrews and moles.

Description.— Body cylindrical; head tapering to a pointed snout; fore-limbs short, with large feet, terminated with strong flat claws; eyes very small, and covered with fur; ears merely small orifices; fur soft like velvet.

American Mole, or Shrew Mole (*Scalops Aquaticus*). Great care must be taken to avoid confounding this animal with its European namesake (*talpa*), of Cuvier, to which it is very similar.

The adaptation of the structure of animals to their modes of life, is perhaps in no instance more apparent than in the organization of this creature. Its short and strong fore-limbs, broad, firm feet, and powerful claws, pointed nose, of which the extremity moves in all directions, the round form of its body, and minute eyes, are all so befitting the place and manner of its life, that without the combination of these parts, it could never exist. Its eyes are adapted to the mere perception of light, since distant vision would be useless to one living entirely under ground, and being so densely covered with a silky fur, are proof against the moist earth, through which it travels. Its sense of hearing is very acute, diving into the earth with a facility perfectly astonishing.

In the construction of its dwelling it displays much taste and judgment. This consists of a little hillock in some dry place, from which paths run in all directions, each terminating at the surface, where a small aperture is left. These paths, as well as the ground about its headquarters, are made solid by the continual passing of their inmates, so that they not only may not admit water during rainy weather, but serve also as a means by which they obtain their daily food, consisting of worms and insects, which finding their way into them cannot escape, and thus fall an easy prey.

All attempts at taming a mole have hitherto proved unsuccessful; we, however, subjoin the following account of one brought by some young people to the Rev. J. C. Wood, an eminent naturalist. It ran about in a large box in which it was secured, with great agility, thrusting its long and flexible snout into every crevice. A little earth was placed in the box, which it entered and re-entered, scattering it tolerably evenly here and there, twitching every now and then, with a quick convulsive shaking, the loose earth from its fur. It was unremitting in its efforts to get through the box, but the wood was too tough for it to make an impression; and after satisfying itself that it could not get through a deal board, it took to attempts to scramble over the sides, ever slipping sideways, and coming down on its forefeet. Its sight and smell seemed to be practically obsolete, for a worm placed close to its nose was not detected; but no sooner did it touch one than in a moment it flung itself upon it shaking it backward and forward, till, getting it fairly into its mouth, it devoured it with a greedy crunching sound.

Having heard from popular report that a twelve hours' fast would kill a mole, Mr. Wood resolved to try the experiment, so having dug a handful of worms he placed them in the box. In its movements backwards and forwards it came upon this mass of worms, on which it flung itself in a paroxysm of excitement, pulling them about in every direction; at last having settled on one, it commenced operations, the rest making their escape to the loose mould. Thinking it had now a sufficient supply, two dozen worms having been put in, Mr. Wood shut up the box, which was not opened until the next morning. Twelve hours had elapsed since the supply was inserted, but as it must have spent an hour in hunting for and devouring the others, eleven hours probably had only gone by since the last worm was consumed, but the mole was dead.

The extreme voracity and restless movements here recorded show its value to the agriculturist; for though generally considered a perfect nuisance in gardens and lawns, yet its destruction of worms and grubs might still show a balance in its favor: and in certain localities, such as old rocky pastures, by throwing up and loosening the soil, and as a subsoil drainer who works without wages, it is of great benefit.

There is another species of this family, much more rare, the Star-nosed Mole (*Condylura cristata*). This mole has a slender elongated muzzle, terminating in a vertical circular disk of from eighteen to twenty cartilaginous fibres. When in confinement, these tendrils or fibres are in perpetual motion. Its geographical limits are not yet established, but it is known from Hudson Bay to Virginia. It is found about old buildings, fences, and stone walls, and occasionally it finds its way into cellars, where, if there is a shallow vessel containing water or milk, it will be sure to terminate its existence from its inability to escape, through clumsiness. All this family pass their winters in a state of torpidity.

The Shrew Mouse (*Sorex*) is remarkable for its diminutive size and apparent helplessness, rarely showing itself by day. Measuring only 2½ to 5 inches, it may properly be considered the smallest mammiferous animal belonging to this continent. Although cats will destroy these little creatures with as great eagerness as they do mice, it is a well-ascertained fact that they will never devour them, probably from the strong musky smell they emit. They frequent the long grass in orchards, and the outskirts of gardens.

There are several species, viz:

SOREX DEKAYI, dark slate blue, 5 to 6 inches long. Not common.

S. BREVICAUDUS, the short tail shrew, furlong, head large, color blackish lead, length 4 inches, very rare.

S. PARVUS, brownish ash color, feet flesh-colored, length 2½ to 3 inches.

S. CAROLINENSIS, iron gray, 4 inches in length, very little known.

CHAPTER II.

ORDER CARNIVORA—A CHAPTER ON BEARS—THE BLACK BEAR—A NIGHT'S SPORT IN LOUISIANA—THE GRIZZLY BEAR—CAPTAIN MARCY'S DESCRIPTION OF IT—THE POLAR BEAR—ITS HABITS AND PECULIARITIES.

The family next in order is the CARNIVORA, or flesh devouring. They fulfil their destined office in the scheme of creation by checking excess in the progress of life, and thus maintaining, as it were, the balance of power in the animal world.

They are characterised by having six conical front teeth in each jaw—the molars formed for cutting and tearing, rather than grinding. Of these, the Bears will engross our first attention.

Description.—Teeth adapted for either flesh or vegetable food; limbs thick and stout; gait heavy and sluggish; feet broad; head large; tail very short.

There are only three species of this animal found here, viz.: the Black, the Grizzly, and the Polar or White Bear, though four are usually described; but the Brown Bear is not to be ranked as an inhabitant of this northern continent; though it has frequently been mentioned by travellers, yet there is abundant reason to believe that they have mistaken the young of the Black Bear, the accounts of their being seen having been confined to the regions where the black or grizzly bear are found. The bear is an animal of great strength and ferocity, passing a great portion of the winter in a state of torpidity and inaction, in dens or hollow trees.

THE BLACK BEAR (*Ursus Americanus*) is peculiar to this country, his range extending from the shores of the Arctic Sea to the southern extremity of the continent; his food principally consists of grapes, wild fruits, the acorns of the

B

live or evergreen oak (on which he grows excessively fat), *larvæ* or the grub worms of insects, insects themselves, and honey, though when pressed by hunger he refuses scarcely anything, his teeth being fitted for a vegetable diet; he seldom attacks other animals unless compelled by necessity; though Major Long, in his explorations in Missouri, saw him " disputing with wolves and buzzards for a share of the carcasses abandoned by the hunters." When he does seize an

animal, he does not, as most others of the *Carnivora* do, first put it to death, but tears it, while struggling, to pieces, and may be said really to eat his victim alive. One distinguishing mark between the European and American Bear is in the latter having one more molar tooth than the former, and also in having the nose and forehead nearly in the same line. It is mostly met with in the remote and mountainous districts, but is becoming more scarce as the population increases. The yellow bear of Carolina is only a variety of this species.

The black bear will not attack a man, but invariably runs from him, unless wounded, or accompanied by its young, when, if molested, it fights very savagely. The old story of the bear sucking its paws, to derive nourishment therefrom when hungry, has doubtless arisen from the slow circulation of the blood in the extremities for several days after recovering from its winter's sleep, which creates an irritation in the paws, alleviated by sucking them, just as we see a dog licking its feet when pierced or lacerated by a thorn.

Bear hunting by moonlight in the Southern States is a favorite amusement, especially in Louisiana. The writer remembers a night expedition of the kind, sallying forth from the hospitable mansion of Major H—, on the Bayou

Goulard, about a hundred miles north of New Orleans. For several nights great depredations had been committed in a large maize plantation some ten or twelve miles distant, supposed to be the work of wild cattle, a few of which had been seen in the neighboring swamps and cane-brakes. A party was at once formed to stay the mischief. It was a lovely cloudless night as we reached the plantation, the moon shining out in all her splendor, and the rich perfume from a magnolia swamp in the vicinity hung upon the breeze as it only can hang in the South. After seeing the rifles all prepared, each member of our party, in eager anticipation, arranged themselves around the spot, preparing to surround the depredators after they entered, and so make sure of them on their retreat. Scarcely had we taken our positions, when a rustling among and waving of the maize showed the approach of one or more animals. Two shots were fired in quick succession by Major H—, followed by several others from different quarters, and three fine black bears were measuring their length upon the sward, whilst two others had escaped in the general confusion. They were covered with fat; and we learned afterwards that several plantations higher up the Bayou had been entirely laid waste, doubtless by the same marauders.

The Grizzly Bear (*Ursus horribilis*), grissly *gray*, grizzly *horrid*, is the largest and most ferocious of its kind, as well as the strongest and most formidable animal of this continent. The name was given to it by Mackenzie, in 1801; nothing satisfactory was known of it till the exploring party of Lewis and Clark in Oregon in 1805 met with it frequently, and left it in the hands of Say to describe scientifically. The description of it given in the *Jardin des Plantes*, in Paris, states that "it combines the ferocity of the jaguar with the courage of the tiger and the strength of the lion."

This bear is no less capable than the other species of subsisting on vegetables; but the supposition of hunters, that it is wholly carnivorous, is easily accounted for, seeing it shows so uniform a ferocity in destroying the life of any animal that falls into its power. It inhabits the country adjacent to the east side of the Rocky Mountains, not extending further south than the confines of Mexico, and affords a very good example of the limitation of species. It has been suggested by Godman, that this animal once inhabited the Atlantic States; but no remains of it have been found to prove this, and he merely starts the idea from a tradition among the tribe of Delaware Indians, that "a big naked bear" (that, certainly, does not correspond with Bruin in question) existed formerly on the banks of the Hudson River.

Although contriving, sometimes, to ascend old leaning trees in search of honey, he cannot ascend perpendicularly small trees, as is shown by the numerous statements of travellers who, when pursued, have climbed a tree, where they have remained many hours, the terrible beast keeping watch below, and shewing signs of rage because unable to reach his prey. So much are their powers respected by the Indians, that they consider it a feat next to that of taking the scalp of an enemy, to kill one of them. Their strength is so great, that they have been seen to kill a large bison, and seizing him with their teeth, drag him up a steep hill. When full grown and fat, they sometimes weigh as much as 1,800 lbs.

Old Adams, or as he was better known under the euphonious title of "Grizzly Adams," devoted many years of his life to the trapping and taming a number of these animals; the account of his hair-breadth escapes is full of the most marvellous exploits; but the result in the number of Grizzly Bears in subjection, in his menagerie, being chained only to posts, and not in cages, shows what man can do by toil and perseverance.

The following notes of the habits and character of this animal were transmitted to the writer, by Capt. R. B. Marcy, U. S. A., whose name is well known among men of science:

"This bear is, in some respects, the most sagacious animal I have ever met with. Before lying down, he goes several hundred yards in the direction from which the wind comes, then turns around and goes back some distance with the wind, but at a short distance from the first track, after which he makes his bed and lies down. Should an enemy now come upon his track, he must approach him with the wind, and with the bear's keen sense of smell he is certain of being made aware of the approach before he is himself seen, and thus is enabled to make his escape.

"When pursued, the grizzly bear sometimes takes refuge in a cave, and the hunters then endeavor, by making a dense smoke at the entrance, to drive him out; but instead, he frequently, when the smoke becomes too oppressive, approaches the fire, and with his fore paws beats upon it until it is extinguished, then returns into the back part of the cave."

"An anecdote was related to me by a Delaware Indian, which goes far to prove this curious animal one of the most stupid in the brute creation. He says, that when the bear enters a cave, it sometimes becomes necessary for the hunter to take his rifle, and with a torch to guide him, follow Bruin in. One would imagine this a very hazardous undertaking, and that the bear would soon eject the intruder; but on the contrary, he sits upright upon his haunches, and with his forepaws covers his face and eyes until the light is removed. In this way the hunter is enabled to approach very close without danger, and taking deadly aim with his trusty rifle, poor Bruin is no more.

"As a set off, however, to this stupidity, an acquaintance of mine, an old bear hunter from the frontier of Texas, removed to California, and shortly after his arrival there, went out to hunt a "Grizzly," and followed one of them into a dense thicket, where he came upon him and gave him the contents of his rifle. No sooner had he done this, however, than the bear turned upon him, and in a few minutes literally tore him in pieces."

THE POLAR BEAR (*Ursus maritimus*) is the next species, met with far up among the icebergs of the Arctic Seas; it is peculiar to those regions, being found only along the sea coasts of the North, where it is so common that no voyager returns without being able to give more or less vivid or frightful accounts of its power or ferocity. The color of its fur is a silvery white, tinged with a slight yellow hue, similar to the creamy yellow which edges the ermine's fur. Its head is so small and sharp-pointed, that there is a very snakelike aspect about that portion of the animal's person. And this shape of the head is the more remarkable, for whereas in other bears the muzzle is separated from the forehead by a well-marked depression, in the polar bear, the line from the forehead to the nose is almost continuous. The sole of its foot is covered with a thick fur, intended, doubtless, for the double purpose of protecting the extremities from the intense cold, and of enabling the creature to tread firmly on the hard and slippery ice.

To most other animals, cold is distressing; to him it is welcome and delightful. In captivity, he seems to suffer much from heat, and his restlessness, from this cause, can only be quieted by keeping him supplied with a water-tank, or by throwing repeated pails of water over him. He is a capital swimmer, catching seals in the water, and diving in search of fish, when not otherwise satisfied. He is often found miles from land, floating on the Arctic ice, from which he swims to the shore without difficulty. In the walrus he meets with a fierce enemy; the dreadful combats that occur, at times, between them, generally terminating in the defeat of the bear. Probably in consequence of the extreme cold which prevails in the high latitudes it frequents, and the absence of vegetation in its haunts, its food is almost entirely of an animal nature, consisting of seals, fish, and the carcasses of whales, though it is able to live exclusively on vegetable food, as has been proved by experiments.

Its fore-paws are frequently rubbed bare, which is thus accounted for: to surprise a seal, a bear crouches down with his fore paws doubled under him, and pushes himself forward with his hind legs, till within a few yards, when he springs on his victim, either in the water or on the ice. When engaged in the pursuit of seals as they are lying on a rock or ice-raft, it employs a very cunning mode of approach. Marking the position in which its intended prey lies, it dives, swimming in the intended direction, only approaching the surface to breathe, finally ascending in close proximity to the slumbering seal, whose fate is now settled; for it cannot take refuge in the water without falling into the clutches of its pursuer, and if it endeavors to escape landwards, it is speedily overtaken and destroyed by the swifter-footed bear.

Its capabilities of scent are wonderfully acute, for it will find out, by that sense alone, the little breathing holes which the seals have made through the ice, even though covered with a uniform coating of snow. Even the Esquimaux dog, specially trained for this very purpose, is sometimes baffled by the difficulty of discovering so small an aperture under such circumstances, which nevertheless is no obstacle to the bear.

The Greenlanders never eat the heart or the liver, saying that these parts cause sickness. It is a curious fact, that the liver of this animal is, to a certain extent, poisonous, causing painful and even dangerous symptoms, to those who have partaken of them,—a circumstance unknown in almost every other animal; for the liver of the black bear when dressed on skewers, with alternate slices of fat (Kabob fashion) is esteemed a luxury by hunters. This fact was noticed by Barentz, who nearly lost three sailors from eating it, and it has since been verified by Capt. Ross. .

CHAPTER III.

THE RACOON AND ITS CHARACTERISTICS—THE BADGER, THE WOL-
VERENE, AND THE GLUTTON—THE WEASEL TRIBE—THE SKUNK,
THE FISHER, MINK, SABLE, WEASEL—DESCRIPTION OF A SABLE
LINE—THE OTTER AND ITS HABITS.

There are few parts of North America, in which the RACOON (*Procyon lotor*) has not been found.

Description.—Head short with fox-like appearance; ears small; tail long and bushy; muzzle tapering, projecting considerably beyond the mouth; color brownish, with broad black patch across the eyes, margined with white; nocturnal.

This animal has been quaintly described as having the limbs of a bear, the body of a badger, the head of a fox, the nose of a dog, the tail of a cat, and sharp claws by means of which it climbs trees like a monkey. This combination may have given rise

to the expression "a queer 'coon." The circumstance which has procured for it the name *lotor* is very remarkable; it is the habit it possesses of plunging its food into water, as if for the purpose of soaking or cleansing it. Some naturalists have supposed it to be not so liberally supplied with salivary glands as most animals, but there is no conclusive proof of this. From its fondness for water it is usually found in low wooded swamps, making its lair in some hollow tree. It is nocturnal, restless, and mischievous in its habits, feeding on wild and domesticated fowls, frogs, lizards, fish, and insects. The tail of the Racoon is never affected by even the coldest weather; hence, it never gnaws it, as other animals of its species are known to do, especially the *Coati* of South America, of which the most marvellous accounts have been given, that it devours its own tail. This however has doubtless arisen from the extreme length of that appendage, in which the blood circulates feebly, thus exposing it to the slightest influence of cold or frost; the irritation thereby produced, leading the animal to gnaw and scratch its extremity to allay that irritation, till it not unfrequently falls a victim to spinal disease produced by this expedient. The Racoon is easily susceptible of domestication; one formerly in possession of the writer being as tame as a cat, and sitting up on its haunches to receive its food in its forepaws before devouring it, and being remarkably cleanly in its habits. Occasionally it commits great depredations among the fields of Indian corn while in the milky state; and this, together with its occasional descents upon the barnyard, scarcely compensates the farmer for its zeal in digging up and devouring grubs or the larvæ of injurious insects.

THE AMERICAN BADGER (*Meles Labradoria*) has only recently been ascertained to be a distinct species from the European; it was formerly looked upon as a new variety, till the publication of Sabine's Appendix to Long's Expedition.

Description.—Color hoary with a white stripe down the forehead, body robust, long on the legs; ears short and wide. The old stories of the life of the Badger being gloomy and wretched from its underground habits, are ridiculous, for Nature evidently destined it for a subterranean and solitary life. It is entirely inoffensive, and being like the racoon, nocturnal, little is accurately known respecting it. The

American species has a short tail and long claws, which are of a light horn color: the European, on the contrary, has a longish tail, and short claws, nearly black. It is found, in the greatest abundance, in the plains adjacent to the Missouri and Columbia rivers, and in Oregon, but individuals are met with here and there all over the continent.

The WOLVERINE or GLUTTON (*Gulo luscus*) is common to both the Old World and the New.

Description.—Body long and low on the legs; fur loose and shaggy; tail very bushy, covered with long pendulous hairs,—frequently confounded with the Bay Lynx (*Felis rufa*), whose habits conform much more to the stories in existence attributed to our wolverine. The statement that it ascends trees for the purpose of leaping down upon the necks of passing animals, and that it takes up with it certain moss of which deer are fond, dropping it immediately under the tree to entice them, has been so frequently repeated that it is generally looked upon as a fact, though the authorities originating these accounts, give nothing as proof more satisfactory than hearsay. A well known American naturalist remarks under this heading " the necessity of scepticism becomes obvious." It inhabits the northern part of America generally, but is everywhere a rare species. Professor Emmons states they still exist in the Hoosac Mountains of Massachusetts. Very little, however, is known accurately respecting it or its habits.

Few, if any, among the small quadrupeds of this continent, equal in beauty the family of *Mustelæ* or Martens, of which the Skunk, the Mink, and the Ermine are best known.

Description.—Long vermiform bodies on short feet; neck long; ears short and rounded; tail long, rarely bushy.

One peculiarity of this species is, that when pursuing their prey, they resemble hounds running on a trail, with tail erect and following by scent. THE SKUNK (*Mephitis Americana*) is well-known and detested everywhere throughout the country. Its peculiar organs of self-defence render it, however, highly interesting to the naturalist : these are, a most fetid discharge, sickening in the extreme, and most difficult to get rid of,—not proceeding from the bladder, as it is usually thought, nor distributed by its tail over its enemies, as has been supposed,—but which is ejected at will by muscular exertion from two glands at the root of that organ, which it at the same time elevates, in order to prevent it coming in contact with the detestable matter, which must be as injurious to itself, as to its enemies. These discharges, at night are said to be luminous. It is a curious circumstance that it never makes use of this provision of nature, unless attacked by a larger animal than itself. It is altogether nocturnal, being most active just after evening closes in, or immediately before day-break. It generally makes its own burrow, feeds on birds and their eggs, frogs, field-mice, and other small quadrupeds. Its fur is coarse, and of no value.

The following interesting sketch of the skunk is taken from Godman. " Persons called by business or pleasure to ramble through the country during the morning or evening twilight, occasionally see a small and pretty animal a short distance before them in the path, scampering forward without appearing much alarmed, and advancing in a zig-zag or somewhat serpentine direction. Experienced persons generally delay long enough to allow this unwelcome fellow traveller to withdraw from their path ; but it often happens that a view of the animal arouses the ardor of the observer, who, in his fondness for sport, thinks not of any result but that of securing his prize. It would be more prudent to rest content with pelting this quadruped from a safe distance, or to drive it away by shouting loudly ; but almost all inexperienced persons, the first time such an opportunity occurs, rush forward with intent to run the animal down. This appears to be an easy task ; in a few moments it is almost overtaken ; a few more strides and the victim may be grasped by its long and waving tail, but that tail is now suddenly curled over the back, its pace is slackened, and in one instant the condition of things is entirely reversed; the lately triumphant pursuer is eagerly flying from his intended prize, involved in an atmosphere of stench, too stifling to be endured."

The FISHER (*Mustela Canadensis*), although twenty years ago numerous, is now becoming scarce. It is known and described also under the title of " Pennant's Marten :" but among the many inaccuracies common to ordinary works on Natural History, is its name " the Fisher;" for this would lead one to infer that its habits are aquatic. Hearne, however, states that it manifests as much repugnance to water as a cat. It is said to have received this appellation from its fondness for the fish used for baiting traps. The early hunters about Lake Oneida were in the habit of soaking their fish over night, and leaving it to drain preparatory to using it; this was frequently carried off by the gentleman in question, whose tracks were plainly seen around, and it has, like the wolverine, been known to follow a " sable-line," destroying twelve out of thirteen traps in one night, in a trail fourteen miles long. It climbs trees easily, living in their hollow trunks, and prefers marshy, woody swamps, near watercourses and lakes. It is not unlike the European polecat.

The SABLE (*Mustela Martis*) is a very active, pretty little animal, inhabiting the elevated woody districts of the North: it is very scarce wherever civilization extends, but was seen abundantly in Oregon, by Lewis and Clarke. It has never been known to become domesticated. It takes up its quarters in trees, and is very carnivorous, living principally upon squirrels. Hunters state that the further North it is met with, the darker is its fur; they also affirm that in the beech-nut season it will never touch bait, carefully avoiding their traps, and that it becomes excessively fat at this time; we may however conclude that it does not use the beech or other nuts as food, but probably fattens itself on the number of small quadrupeds which are congregated together more thickly than usual, to feed on the mast. In the Hudson Bay territory a line of traps will be set for it called a "a sable line," sometimes sixty or seventy miles in length, at the rate of from six to ten a mile, visited by the trappers perhaps once in a fortnight. These traps are very simple, being generally made of long chips cut from the nearest tree, which driven into the ground form three sides of a square about six inches across; the bait is then placed on a stick laid crossways between the main support and prop of a heavy log or rough board, which falls the moment the bait is touched, crushing all under it; the top is then covered with some boughs of spruce or hemlock thrown lightly over it, and left to do its silent work. Fishers and wolverines will follow one of these sable-lines, breaking into the traps from behind, and destroy the bait as well as the captive if any is there. The American sable has been often confounded with, but is quite distinct from the pine marten of Europe.

The SMALL WEASEL (*Mustela Pusilla*) is supposed by some to be, and on the authority of Bonaparte is, the ermine in its summer coat, but this is very doubtful. It is very voracious and very tenacious of life. It is common about old walls, farm buildings, thickets near lonely houses, &c. It must not be confounded with the

ERMINE (*Putorius Erminea*). This weasel is very destructive to poultry, but its injuries are perhaps counterbalanced by the numbers of mice and rats it destroys in barns, stacks, and about the farm buildings. It is very active, nocturnal in its habits, and frequents wood-piles; in its white winter coat, with tail tipped with black, it is sometimes called the Catamingo, or White Weasel.

The last of the Weasel family we shall describe is the MINK (*Putorius Vison*). Its name is corrupted from the word *Mœnk*, given by the early Swedish settlers in the United States. It is well known, and is met with in all parts of the country, frequenting the banks of streams and swampy ground. In the West there is scarcely a stream on the banks of which its footprints are not visible; it feeds on fish, fresh-water shell-fish, and is closely allied to the otter in many of its habits; it can remain a long time under water, either when pursued or when searching for food. An odor is said to be emitted by it when attacked, somewhat between that of a cat and a skunk; when closely pressed it sets its pursuer at bay, arching its back like a cat, snarling and turning with the greatest rapidity, and makes a desperate resistance before it is captured.

To a casual observer the *Mustelidæ* would seem very scarce; but as night is the season for their operations, they seldom or never shew themselves by day; their habitat may be frequently passed by unwittingly, except when winter reveals it by their trail in the snow. In the woods and rocky regions of the West and Hudson Bay they are most numerous; but enough are left everywhere for them not to be classed among the rarer animals.

The OTTER (*Lutra Canadensis*) was long confounded with its European congener, till proved by Sabine to be distinct. It is found throughout the whole continent.

Description.—Amphibious; broad palmate feet; tail more or less horizontally flattened; head broad and rounded; blunt muzzle; ears very short; eyes small.

The otter feeds exclusively on fish, and aquatic animals, though in a state of domestication it will devour raw meat. It fights with great fierceness, and is more than a match for a common sized dog. Its legs are very short, and its feet webbed, and better fitted for swimming than for running upon land; being so eminently aquatic in its habits, it is seldom seen far from the water.

The otter is becoming scarce as the country is being cleared up; and is, like the Indian, compelled to give way before the approach of man, retiring further westward and

northward yearly. In places where it used to be most abundant, no trace of it is now found, except in the names of streams or localities, such as Otter Creek, Otterville, &c. It is very sagacious and wary; its fur ranks next in quality to that of the beaver, and is greatly used in the manufacture of hats. The otter is too wary to touch baited traps; they are accordingly placed in the water at the foot of their slides, for which they have a curious fondness. These slides are thus formed: a number of them (for they live frequently in families like the beaver) will select a spot where the river bank is clayey, and having rendered it smooth by removing sticks, stones, &c., they start from the top, one after another, with a velocity that brings them plump into the water. Major Long thus jocosely alludes to them: "These slides are sometimes borrowed by boys bathing; who, however, not recollecting that the otter is protected by a thick fur against friction, find that notwithstanding the apparent smoothness, the fine sand in the clay has robbed them of a broad surface of cuticle, and that an otter slide is not altogether suited for human recreation." The otter can be domesticated like the beaver, and becomes very docile.

There is another species, (*Lutra destructor,*) so called from its destroying the beaver dams and houses, probably in search of the young beaver. It is met with in the Hudson Bay territory, but together with the third species, *Californica*, of the Pacific coast, little is accurately known of them. The Ojibbeways, however, knew long ago of their existence, from the two different names used for the two species in their language. The sea-otter is exclusively resident within the 49th and 60th degrees north latitude.

CHAPTER IV.

THE DOG—THE ESQUIMAUX OR ARCTIC DOG, ITS HABITS &c., AND FACTS RELATIVE TO IT, RECORDED BY RAE AND KANE: ITS USE IN SLEIGHING—THE FOX, HABITS, AND ANECDOTE OF ITS CUNNING, OR FOXINESS.

Of the American dog there are supposed to be eight species indigenous, though this is as yet an open question; the wolf, the fox, and the jackal, being each claimed as the originator of the species, in different countries.

Description.—Six cutting teeth in each jaw,—canine teeth four, one on each side of either jaw;—tongue, soft; five toes on the fore-feet, four on the hind feet: they never sweat; drink by lapping.

The *Lagopus*, a native of Greenland and Spitzbergen, is supposed to be the true originator. In this animal, the Arctic dog, we find an illustration of the alteration of species in connection with civilization, not only in its variety of form, but from the established fact, that the Esquimaux dogs had never been known to bark until they heard their domestic cousins, which accompanied the discovery ships of Arctic expeditions, giving tongue, and so, by imitation, acquired the habit, now as common to them as to our canine followers.

The conquest of the dog is the most complete, singular, and useful ever made from the animal kingdom by man. The whole species has become his property; each individual is devoted to his master; assumes his manners, knows and defends his property, and remains his true friend till death; and all this from the purest friendship, and even in spite of starvation and cruelty. Of all animals, this is the only one which has followed man in every condition through all the regions of the globe, and been his defence against the prowling beasts of the forest and the desert. But as so much has been, and is continually being written on the subject, and as every one knows numerous anecdotes connected with this animal, we will confine ourselves strictly to the true American or Esquimaux dog: He is large and powerful, equalling the mastiff in size; hair long and thick, tail long and bushy, and turned over the back; ears short, pointed, and erect. And here, speaking of his tail being turned over his back, let us mention that the domestic dog is distinguished from the other species of this tribe, by his *recurved* tail,—this member in the others being straight. This is the dog which draws sleighs, or sledges in Arctic phrase, and transports loads from place to place, with one or more persons in them, over the frozen

snows. He is good-tempered and very enduring, and though often cruelly treated, is still willing to do everything in his power at the command of his master. What the camel is to the Arabians, and the reindeer to the Laplanders, the Esquimaux dog is to the inhabitants of the Arctic regions. These creatures seem designed to work in the harness, and hence, it is said, perform their duty almost instinctively, requiring but little training or breaking in. The sledges are usually constructed for only a single person, and are drawn sometimes by three, but more frequently by five dogs, one of which acts as leader. They are guided not by reins, but by striking on the ice with a stick, the voice being occasionally employed; and in a country where there are no roads, the direction must depend on the instant obedience of the leader to the indications of the driver, otherwise danger would often be incurred from a precipice or impediment. When any of the dogs are inattentive to their duty, the rider punishes the delinquent by throwing his stick at him, which he dexterously again picks up without stopping. It is said, these cunning animals very soon ascertain when the stick is lost; and unless the leader is uncommonly well trained, the driver is in peril, since they set off at full speed, and do not stop till they are exhausted, or the sledge overturned. They possess the most wonderful sagacity in finding their way during snow storms, when their master can see no path, nor even keep his eyes open in the blinding storm. In such cases they seldom miss their way; but if at a loss, they will go in different directions, until satisfied of the course, probably by the smell. If during a long journey, it is found that the place of destination cannot be reached, and it is impossible to proceed further, then the dogs are unharnessed, and lying down in the snow with their master in the midst, they keep him from freezing, and if necessary defend him from danger. A popular writer and traveller, Bayard Taylor, says, that "driving Esquimaux dogs is very much like driving a lively sturgeon in rough water. As soon as you are seated in your sledge, which is like a little canoe, off they start, and as the bottom of the sledge is perfectly round and slippery, it is no easy matter to maintain your balance. If you are a new hand, your first experience is head-first downward in a snow-drift." The value and use of the Esquimaux dog in the Arctic expeditions, seem to have been appreciated only by our recent explorers, Kane and Rae; both of whom made frequent use of them, in scouring those inhospitable wastes in search of the missing Franklin and his crew; had *he* been provided with those necessary appendages of Arctic travel, we should not have the mournful detail recorded by the natives to Dr. Rae in 1854, " that a band of forty white men *dragging their sledges* along the coast of King William's land were making apparently for the great Fish River; that all, even with one who seemed to be an officer, were *dragging on the haul ropes* of the sledge." Both these explorers speak in the highest terms of the assistance these dogs afforded to their party; and from Dr. Kane the writer gleaned what knowledge he has of their habits. Snow he stated to be their substitute for water; and on a lump of it, or ice given to those he brought to New York with him, they would roll with the greatest delight. The snow he observed they did not *lick* up, but by repeatedly pressing with the nose, they would obtain a small lump or ball of it, which they then drew into the mouth with their tongue.

The following account is given of the habits and disposition of one of these dogs by its owner: " Even if coaxed and fed by a stranger, he had so strong an attachment to his master, that he would merely take the food without returning thanks either by looks or wag of the tail. He never barked, and would snap at those he did not like, without a growl or the least notice. He was remarkably cunning, resembling in that respect the fox, for he was in the habit of strewing his meat around him to induce fowls or rats to come within his reach, while he lay watching, but pretending to be asleep, and when near enough he would pounce upon them, never missing his aim."

THE FOX (*Canis Vulpes*), when compared with the dog family, is found to be lower in height, in proportion to its length.

Description.—Its nose is sharp, limbs slender, tail bushy and long, reaching to the ground. This family, generally speaking, lead nocturnal lives, and have a propensity to burrow in the earth, which dogs never do; in habits they are unsociable, never, although capable of being tamed, becoming truly domestic; they are sly, cautious, and " cunning as a fox," being ever ready to destroy all such animals, especially young and tender ones, as they can master. When caught in a trap they will sacrifice the limb, by gnawing it off, and thus escape. There are five species ascertained to be peculiar to this country, though Geoffrey adds a sixth, since, however, ascertained to be only a variety of the black: of these THE RED FOX (*V. fulvus*) is by far the most common. This has been thought to be identical with the common fox of Europe—but the fineness of its fur, the brightness of color, slenderness of body, and the form of its skull, clearly prove it a distinct species.

THE GRAY FOX (*V. Virginianus*) is very common, being found more in the vicinity of farm buildings than the red one. It is preferred by the hunters, since it does not start off directly from its haunts, but after sundry doublings is generally captured near its starting point. THE BLACK OR SILVER FOX (*V. Argentatus*) is found through the northern-most parts of the Continent, as well as in Asia, but is very rare, and its skin is accounted one of the most valuable furs. THE SWIFT OR BURROWING FOX (*V. Velox*) inhabits the Missouri or the Rocky Mountains, and always burrows; hence its name. Its swiftness is inconceivable, outstripping the antelope, and may be compared more to the flight of a bird. The notes respecting it taken by Say, were lost, and as no other naturalist has given an accurate description of it *from observation*, very little can be said of its habits. The fifth kind is THE ARCTIC FOX (*Canis Lagopus*). This frequents the higher latitudes, and only comes a few degrees below the Polar Circle. It is captured to a great extent in the Hudson Bay territory; is very voracious, as a proof of which Capt. Lyon, who accompanied Parry, mentions having found in the stomach of one which he examined " a mass of rope-yarn and line, among which some plaited pieces were fully six inches long." It is very cleanly, and no unpleasant smell is preceptible from it—an exception unknown to the rest of the species. It is of a pure white in winter, becoming brownish or gray in summer.

The following anecdote is given of the gray fox: A few years since, one was started in New Jersey, and after running a few miles before the dogs, was shot at and apparently struck, as he made several somersaults and then fell, but recovering, started off again. Another hunter next had a chance, and poor Reynard again fell, was taken up and carried home to all appearance a dead fox, and accordingly thrown into a corner of the room. While the hunters were at supper, the supposed dead animal was seen to raise himself on his fore-legs, cautiously looking about to see what chance there was of escape, but finding himself observed, he again resumed the quiescent state. One of the party now passed a piece of burning paper under his nose, but to all appearance he lay senseless as a stone. The room, however, was closed for the night, and it was found in the morning running about inside as though nothing had happened. On examination, not a bone was found broken, and with the exception of a slight wound in the shoulder and a soiled coat, he was as well as ever.

The writer some years ago obtained a cub of the *V. Fulvus* about three weeks old, which from the care bestowed upon it, became as great a favorite with his children as a dog, evincing with them no savageness, but whining and snapping at any stranger. Though chained in the garden beside an artificial burrow, near the resort of his owner's poultry, amongst whom he had been reared, he never attempted to touch one of them; but should some luckless chicken from a neighboring yard find its way into the limits of his tether, it never returned to tell the tale.

CHAPTER V.

THE WOLF:—DESCRIPTION, HABITS AND RAVENOUS CHARACTER. ITS DEPREDATIONS: INCIDENT AT BIDDEFORD.—THE PRAIRIE WOLF, ITS HABITAT—ATTEMPT TO CAPTURE AND FAILURE—DUSKY, AND BLACK WOLF.

The Wolf. *Description.*—Head broad, muzzle pointed, eyes small; ears erect and pointed, tail long, straight and bushy, tipped with black, voice howling, color various, mostly red, sometimes black or gray.

There is no animal whose character in general estimation is worse than that of the wolf (*Canis Lupus*). And yet when we take into account that he is a universal outcast and entirely dependent upon rapine for his subsistence, we cannot blame him for living as he does, since he must either destroy or starve. The carnivorous tribes are evidently designed for the destruction of others, their teeth and claws being given them for this purpose. On the contrary, herbivorous animals, as the cow and sheep, require no such means of procuring their food, and accordingly are furnished only with teeth for cropping and grinding vegetables. Now although we are bound to protect ourselves from the fangs of the tiger and the cunning of the wolf by the destruction of these animals, yet so far as the animals themselves are concerned, the wolf is no more to blame for killing the sheep than the latter is for plucking the grass, because these are the only means by which the Creator intended these different animals to live. Torturing the wolf therefore for having destroyed the lamb, is no more excusable in us, than punishing the lamb because he happened to pluck some plant which we particularly value.

This sullen and unpleasant looking animal, the most ravenous and ferocious that infests the more temperate regions of the earth, in many parts of which he is the terror and scourge, has the general appearance of the dog, with the exception of the tail, which is straight instead of being curved over the back. And yet there is something in the physiology of the wolf, as well as in his gait and manners, which is at once so repulsive and peculiar, that however tame he may apparently be, he never could be mistaken for even the most wild and savage of the true dogs.

Wolves, like dogs, follow by the scent; and when the prey is too powerful for a single one they combine in packs, and like well trained hounds keep up the chase to the certain destruction of their victim. But this combination of forces never arises from any social or friendly disposition, but only to assist each other in a work of destruction which they cannot perform alone. The moment therefore the object is attained, they attack each other with the most savage ferocity, no one allowing the other, willingly, the least share of the booty to which they all have an equal right. These quarrels over the flesh of their victims, are said sometimes to continue, until many of the weaker ones are themselves destroyed and then devoured by the stronger. Wolves usually select a young or injured deer, and trust more to tire him down, than overtake him by superior speed.

In the summer their prey escapes easily by taking to the water, but in winter the same instinct leads to its immediate capture, for on the ice it is quickly overtaken by its pursuer, and towards spring there is scarcely a Northern lake in the woods that has not numerous carcasses of deer on its frozen surface.

When met with singly, the wolf is a great coward, the American species showing the white feather even more than the European. In the early settlement of this country, THE COMMON WOLF (*Canis occidentalis*) was the common terror and scourge of the farmer, destroying his sheep and his young cattle; but like its European cousin, as civilization extended, so it receded to the remote, wooded, and mountainous districts. In Massachusets and New Hampshire they are still occasionally found, and a few years ago a fine specimen was killed on Talcott Mountain in Connecticut;—every winter in Maine, and throughout the backwoods of Canada we hear of their depredations, though their ancient courage seems to have forsaken them, avoiding the face of man, and confining their attacks to domestic animals,—and that only when pressed with extreme hunger.

In many counties in the States, bounties varying from ten to twenty dollars per head are offered for wolves, paid partly by the State, and partly by the County and Township. The color of the common wolf is various ; mostly pale red, sometimes black or gray. The wolf of Pennsylvania—scarcely now to be met with—is redder than that of Florida, blacker and larger than those found elsewhere ; but they are only varieties and not a distinct species.

The following remarkable incident took place near Biddeford, Maine, a few years ago, and is so well vouched for as to give no doubts as to its authenticity. A resident in that place, a Mr. Adams, was that autumn engaged in felling trees at some distance from his house. His little son eight years old, was in the habit of running out into the fields and woods, and often going where his father was at work. One day after the frost had robbed the trees of their foliage, he left his work sooner than usual and started home. On the edge of the forest he saw a curious pile of leaves, and without stopping to think what had made it, moved the heap, when to his astonishment he found his boy asleep there. Taking him up in his arms, he had scarcely moved many paces before he heard a wolf's distant howl, quickly followed by others, till the woods seemed alive with the dreadful sound. The howl came nearer, and in a few minutes a large gaunt, savage looking wolf leaped into the opening, closely followed by the whole pack. Springing on the pile of leaves, it quickly scattered them in all directions, but finding the bed empty his look of fierceness changed into that of the most abject fear ; for the rest, apparently enraged at being thus baulked of their prey,-fell on him, tore him to pieces, and devoured him on the spot. The victim had probably found the child sleeping, covered him with leaves until he could bring his comrades to the feast ; and unwillingly himself furnished the repast.

THE PRAIRIE WOLF OR CAYOTA (*Canis latrans*) is said by Say to be more numerous than any other on this continent ; its cry very closely resembles the barking of the domestic dog ; and in appearance it so much resembles the Indian dog, that some naturalists believe it to be the original parent of that animal.

On the great western plains of Missouri these animals abound to an incredible extent. They also occur, but more sparingly, in all the north-western portions of this continent, from Hudson Bay to the Rocky Mountains. Their food consists chiefly of rabbits, and such other small animals as live on the prairies. This species also congregates in packs for the purpose of hunting deer and young bisons; and like all others of the wolf family they are extremely cunning and rapacious; not even the fox has more intuitive sagacity in avoiding traps or snares. A member of Major Long's expedition had a strong desire to possess himself of one of these animals alive. He therefore set a trap, of the kind known to boys under the name of "figure four," similar to those used for rabbits, but in size proportionate to that of the animal. First he laid down a plank floor about six feet long, and over this set his trap of the same length, one end being elevated about three feet, like an inclined plane. The bait of meat, attached to what is called the spindle of the trap, was placed in the middle of the floor, and this being touched, the trap would fall and catch the wolf, if indeed he was there.

Now for the result. It is very unlikely that a prairie wolf had ever seen such a trap, and yet, instead of going under and taking the meat, these cunning brutes actually dug a hole beneath the platform, and lifting up the boards, possessed themselves of the bait, sprang the trap, and of course escaped unhurt. A large steel trap, which being trod on catches the foot, was next tried. This was well baited, and covered with leaves. But although their tracks showed that the wolves had visited the place during the night, the bait was untouched. The bait was then hung over the trap, and pieces of meat suspended from limbs of trees in the vicinity in the same manner, so that the trap, being covered with leaves, no one except a wolf could tell under which piece of meat it was placed, and yet in the morning every piece of meat was gone except the one hanging over the trap; various other plans were tried, with the same want of success, until the trapper found that all attempts to catch these animals were useless, and he therefore gave up the trial.

The Dusky Wolf, a variety of the common wolf, frequents also the same region, but is far less numerous, much larger than the Prairie-wolf, and is remarkable for emitting a strong odor.

The Black Wolf, another variety, has been found in British America, but is very rare; it was also seen in the Rocky Mountains by Long's expedition. Desmarest thinks it a distinct species from the European, but not enough is known of it to form any decision.

CHAPTER VI.

CARNIVORA—THEIR HABITS, POWER OF VISION AT NIGHT, AND PECULIARITIES OF THEIR FEET AND WHISKERS.—THE COUGAR, OR PUMA—DESCRIPTION—EARLY RECORD OF—COUGAR KILLED AT SOREL—ANECDOTE OF ITS STRENGTH AND FEROCITY.—THE LYNX AND WILD CAT—DESCRIPTION—DOMESTIC PUSS.

We next come to the Feline family of the Carnivora—the Cat tribe,—of which it will be well to say something before enumerating their species. In it are comprised the most ferocious and bloodthisty of the Mammalia. They hunt chiefly by night, and are exceedingly cunning in the means by which they entrap their victims Their power of seeing in the dark has always been a mystery, nor is it strange that it should be so, since man of all animals has the least nocturnal power of vision.

In all night-prowling animals the eye is peculiarly large, so as to admit a great number of the rays of light, for it is seldom or never perfectly dark in the open air. It was supposed formely that the eyes of cats and owls generated light, their structure being such as to produce a phosphorescence by which objects became visible in the dark! Recent experiments however show, that their extra power of vision is produced by the concentration of the rays of light by the eye of the animal, and that when it is totally dark the eyes of a cat cannot be seen. This faculty then depends on such a structure of the eye, as enables it to collect the scattered rays of light in greater quantities than that of other animals.

In the foot of the cat tribe, the marks of the wisdom of the Creator's design to perform the very purposes for which we see they are employed, are particularly apparent. The power of these animals, so to balance themselves when leaping from a height, as to come down upon their feet, is well known. Even when thrown with the head or back downwards they will turn, so that the feet shall come first to the ground; and from heights which would destroy the lives, or at least break the bones of any other animal, puss will land in safety, and bound away without a limp. This peculiarity is owing to the chief bulk of the foot being composed of elastic tendons, and balls, or cushions, consisting of a substance intermediate between cartilage and tendon, being attached to the sole of each foot, the middle one being made up of five distinct parts, besides a similar pad to each toe. In walking, the cat tribe do not touch the ground with their claws, for these remain sharp even in old age (which some persons may have learnt to their sorrow). But in seizing their prey, or inflicting vengeance, the feet by means of the claws, become instruments capable of holding the victim, or piercing the skin of an enemy. This is performed by an elastic ligament acting as a spring, by which the claw is drawn up or backward, and to bring down which, muscular action is necessary; this is effected by the contraction of a strong muscle to which the tendon is attached, the shortening of which pulls down the claw, attached in its turn by a ligament to the bone.

The long hairs on the upper lips of the cat tribe, are of great importance to these prowlers. They are the organs of touch, each one being connected with the nerves of the lip, so that the slightest contact with any object is known to the animal instantly. These hairs project round the head to such a distance as to equal the diameter of the animal's body, so that by them they can measure the size of an aperture before they attempt to pass through it. The writer well remembers in Leicestershire, when a boy, seeing sundry cats which had been shorn of their whiskers, in consequence of poaching proclivities; the gamekeepers asserting that under such manipulation, a cat would never venture far from home among bushes.

This tribe cannot, like most other of the *Carnivora*, subsist on vegetable food, but must either starve or possess themselves of flesh. Sometimes they drop from a tree, or, lying in wait under cover, they spring upon and secure some unsuspecting animal at a single bound. It is remarkable that some individuals of this tribe are found in nearly every region of the earth; and though the *same kinds* are confined within certain limits, still most parts of the globe appear to be represented by their own peculiar species; thus, the lion and tiger are inhabitants of Africa and Asia, being found nowhere else. In America these are represented by the puma or jaguar, confined to that continent. The caracal is found only in Turkey and Persia; the lynx in Northern America and Canada.

The Cougar (*Felis Concolor*), called also the Puma, Catamount, Panther or Painter (the last term evidently a corruption), and American Lion, is the largest of our species.

Description.—Ears short and distant; no mane; tail long and slender; fur soft and short; color, dark reddish gray.

A great deal of confusion has arisen as to the name panther, which, however, has been decided to be the *Felis pardus* of Linne, an Asiatic animal. The puma was called the American lion by the naturalists who first explored this country; they contended it was a true lion, but *degenerate in size, owing to the climate!* Vanderdenck, in his history of the New Netherlands (now New York State), says: "Although the New Netherlands lie in a fierce climate, and the country in winter seems rather (?) cold, nevertheless lions are found there, but not by the Christians, who have traversed the land without seeing one. It is only known to us by the skins, which are sometimes brought in for sale by the natives. In reply to our enquiries, they say that the lions are found far to the southwest, fifteen to twenty days' journey; that they live in very high mountains, and that the males are too active and fierce to be taken."

The cougar climbs trees with surprising agility; its cry is peculiar, closely resembling the wailing of a child, and in the early settlement of the country has sent a thrill of horror through a whole neighborhood.

The following interesting account of the last recorded capture of the cougar in Canada was courteously handed to the writer by H. R. Gray, Esq., of Montreal, just previous to going to press: " On October 3rd, 1863, as Jacques Gamelin, of La Baie Lavallières, or St. Francis, on the River St. Lawrence, and two other men, were proceeding in a *bateau* to Sorel, C. E., being at the time about four arpents (acres) from land, they were suddenly attacked by a large panther, which swam off and succeeded in laying hold of the boat. Not having fire-arms or other defensive weapons on board, they were obliged to resist the attack with a boat-hook and oars, and after a lengthened struggle, during which the animal displayed great obstinacy combined with strength, he was at last killed, but not before he had actually succeeded in getting his powerful fore-paws on deck; not being able to make his accustomed spring, in consequence of his hind-quarters being under water, he was incapable of injuring either of the men. He was killed and his body sent to Montreal, where it fell into the possession of and was stuffed by Mr. Craig, the well-known taxidermist." Its weight was 131 lbs.,—measured three feet from tip of the nose to the root of the tail.

Hunters and surveyors build a large fire at night, which serves to keep this cautious animal at a distance. Under such circumstances, it will sometimes approach within a few rods of the fire, and they have been thus shot by aiming between the glaring eye-balls, when nothing else was visible. It will seldom attack a person in the day time, unless provoked or wounded. The following account, showing the strength and ferocity of this creature, is taken from Godman : " Two hunters, accompanied by two dogs, went out in quest of game near the Catskill mountains, N. Y. At the foot of a large hill, they agreed to go round it in opposite directions, and when either discharged his rifle, the other was to hasten towards him to aid him in securing the game. Soon after parting, the report of a rifle was heard by one of them, who hastening to the spot, after some search, found nothing but the dog, dreadfully lacerated and dead. He now became much alarmed for the fate of his companion, and while anxiously looking round, was horror struck by the harsh growl of a catamount, which he perceived on a large limb of a tree, crouching upon the body of his friend, and apparently meditating an attack upon himself. Instantly he levelled his rifle at the beast, and was so fortunate as to wound it mortally, when it fell to the ground along with the body of his slaughtered companion. His dog then rushed upon it, but with one blow of his paw it laid the poor creature dead by its side. It was finally despatched with great difficulty." The cougar is fast disappearing; though a severer winter than usual in Canada, brings it about the settlements sufficiently to show what mischief it could be capable of inflicting, if numerous.

THE NORTHERN LYNX (*Felis Canadensis*) is found in both Asia and America.

Description.—Slim in form; ears acute, and more or less tufted; tail short; timid; moves by a succession of leaps, alighting on all four feet at once; color uniform reddish gray.

This is a fierce and subtle animal, but fearful of man. It seldom approaches settlements, preferring the plains and woods of the wilderness. It is found in great abundance in the Hudson Bay territory, and all the northern parts of America; whence as many as eight thousand of its skins have been exported in one year. Its fur is in great demand for muffs, and winter decorations of ladies' garments; it is long and beautifully lustrous, but is commonly colored of a dark brown or nearly black, before it is exposed for sale by the retail dealers. The long sharp tufted ears, and short tail of the lynx give it a peculiar appearance, and distinguish it at once from the rest of the cat tribe.

Its piercing sight has long been proverbial, though its powers in that respect have doubtless been exaggerated; hunters confirm this idea, however, of its discovering objects not visible to other animals. In captivity it is the most vicious of its race, returning the most spiteful menaces in exchange for the kindest treatment.

THE WILD CAT, or BAY LYNX (*Felis rufa*), is peculiar to this Continent.

Description.—Legs disproportionately long to the body, which is slender; ears large, with moderate black tufts; tail slender; color, rufous.

Many species of this Animal have been described under the names *fasciata, aurea, montana, &c.*, but no satisfactory evidence has been given to prove them distinct. The naturalists attached to different expeditions have never been able, either personally or from the Indians, to obtain any other than the one species. It is found, though comparatively rare, in all wild paths of the Northern and the Western States, and Canada.

The origin of the house cat, like that of many other of our domestic animals, has always puzzled those who have undertaken to trace it; nor is it yet settled from what race of the feline tribe it has been derived. Some suppose that the Bay Lynx was its parent; and that domestication for a long series of years has produced the difference between them. But the Bay Lynx is of a uniform grey color, with a short tail, and two or three times as large as our tame cat. The length of the tail, it is true, is no criterion, since we have a tame variety nearly destitute of this appendage; but that domestication should produce a diminution of size is opposed to general observation and experience with respect to other animals. That the cat was a domestic animal among the ancient Egyptians, is proved by feline mummies being found in the catacombs of that people; and as that is sufficient to prove the high and honorable antiquity of the race, we shall leave all further inquiries concerning the genealogy of Puss to those who have more time and patience.

CHAPTER VII.

SEALS — DESCRIPTION AND PECULIARITIES — WHERE FOUND — ESQUIMAUX SEAL HUNTING—THE COMMON SEAL—THE HOODED SEAL—THE GREAT SEAL—THE HARP SEAL—THE FŒTID AND URSINE SEAL.—THE WALRUS.

From very early ages, the family of Seals (*Phocæ*) has been considered a sort of anomaly, bearing the same relation to fish as the bat was supposed to bear to birds.

Description.—Head rounded, no external ears, eyes large, tail very short.

It was doubtless from imperfectly formed observations of these animals that the stories of mermaids, sirens, sea nymphs, &c., originated. In their natural history much remains yet to be elucidated; and this is no matter of surprise when we consider the ignorance which exists concerning mammalia much more within our reach, than these marine inhabitants. Cuvier remarks "we have not the means, except by deduction and analogy, of ascertaining the habits of these half amphibious animals, while procuring their sustenance in the sea." We are, however, acquainted with the physical structure of the animal, and possess a knowledge of its character and habits.

The form of the body of the seal bears a general resemblance to that of a fish. A remarkable formation about them is their having only the extremities of their limbs visible; the remaining part being closely covered by the integument of the body, the fore limbs to the wrist, the hind ones to the heel; the toes are furnished with claws, and are united by a membrane, which serves for the purpose of a fin or paddle. The enclosure by this integument gives the limbs an appearance of extreme shortness, and occasions the animal to crawl on land with great awkwardness and seeming difficulty, to which motion the expressive term "walloping" is applied. In the water they move gracefully and easily. By means of their fore feet they can lay hold of objects with sufficient firmness to drag themselves up shores, and on shoals of ice, however slippery they may be. Even on land their motions are quicker than their appearance would lead one to expect; so that it frequently happens that when they have been dangerously wounded, hunters are unable to overtake them before they get to the water's edge and so escape.

The nostrils of the seal are provided with a peculiar muscular apparatus, by which their orifices are perfectly closed at will, effectually excluding water during submersion. Cuvier states that they are seldom opened, except when desirous of expelling or introducing air into its lungs: and that they then assume a circular form; their respiration is extremely unequal, and performed after long intervals. The quantity of air inhaled seems, however, to compensate for the paucity of the inspirations, for few animals have more natural heat or a greater quantity of blood than the seal. The form of the teeth and the jaws shows them to be carnivorous, their food consisting principally of fish, crabs, and sea birds, which they are able to surprise while swimming.

The females produce two or three young at a time, generally in the winter. When the cubs have acquired strength enough to contend with the waves, the mother conducts them to the water and teaches them to swim about in search of food. When in danger, the safety of her cubs is the chief object of attention with the mother. The male parent, particularly of the ursine seal, seems to take scarcely less delight in the young than the mother. While basking in the sun upon the shore, he eyes them with the greatest complacency, and expresses his satisfaction by licking and kissing them as they sport and tumble about.

All the species of seals live in herds or families along the sea-shore, and are fond of sunning themselves, lying on the beach, rocks, or ice-banks. When they do this in situations in which they are apprehensive of danger, instinct has taught them to take the precaution of posting a sentinel to give alarm, when he sees anything to excite apprehension; besides which, while sleeping, the seal raises its head at frequent intervals, and looks around as if to observe that all is safe within its range of vision. In places where they rarely experience disturbance, they sleep very profoundly and are easily surprised. They are in general very tenacious of life, and survive wounds which would kill most animals; but are much more easily dispatched by blows on the head, especially on the nose, than any other animals.

Seals' are very extensively diffused, for though they seem to prefer cold climates, there is scarcely any sea on the shores of which they are not occasionally to be met with. The old Dutch word "robyn," a seal, is still met with along the northern coast, in such terms as "robin's reef," "robin's head," &c. They are most numerous in the Polar and Arctic regions, furnishing the Esquimaux with food, clothing, and light. Langsdorff, in his voyage round the world, thus speaks of them: " This animal forms such an essential article to the subsistence of the Esquimaux, that it may truly be said of them, they would not know how to live without them. Of its skin they make clothes, carpets, thongs, shoes, and household utensils; nay their canoes are made of a wooden skeleton with the skin of the seal stretched over it, * * * there is no part of the animal that is not turned to some use."

The hunting of seals is consequently prosecuted with great eagerness by the natives of the Polar latitudes. Parry gives an amusing description of the ceremonies attendant on a seal feast among the Esquimaux. He says, "Before cutting the animal up they pour into its mouth, as it lies on its back, a little water, and touch each flipper, and the middle of the belly, with lamp-black and oil. The origin of this ceremony is unknown, but from the seriousness with which it is performed it seems to be one of high importance." The Greenlanders connect strange ideas of honor and glory with the chase of the seal. They will expose themselves to the greatest dangers, wandering over the waves for days together without any other guide than the sun and moon, and as they deem it disgraceful to leave any part of their game behind them, they sometimes overload their limber, crazy boats, and disdaining to save their lives by relinquishing their highly-prized acquisitions, proudly perish with them in the waves.

THE COMMON SEAL (*Phoca vitulina*) is abundant on the shores of the lower St. Lawrence and northern parts of the sea coast generally—being occasionally met with on inland waters connected with the sea. One was taken on the ice in Lake Ontario, near Cape Vincent, in 1823-4, and one was captured in Lake Champlain in 1810, probably having reached there by way of the St. Lawrence and Richelieu river. They may occasionally be seen in winter opposite Montreal in openings in the ice disporting themselves; and a fine young specimen was given to that city by the officers of the "Shandon," taken off the ice in the Gulf in 1864. It is to be regretted that this specimen died, probably from want of a sufficient depth of water, and the too eager curiosity of the people.

THE HOODED SEAL (*Stemmatopus cristatus*) is found on the shores of Newfoundland and Greenland. A fine specimen was captured at East Chester, New York, a few years ago. It occasionally finds it way up the St. Lawrence. This seal has a peculiar appendage to its head, formed by the extension of the skin in front, which can be elevated or depressed at pleasure. No satisfactory idea has been given for the use of this hood, it being supposed by some to be a protection for the eyes, and by others as a reservoir for air when the animal is under water; but DeKay remarks, very truly, "that all speculations, not based on actual examination of an animal in its living state, can lead to nothing else than a mere multiplication of words."

THE GREAT SEAL (*Phoca barbata*) which attains ten or twelve feet in length, is found in the Greenland seas, and northern extremity of America. Very little is known of its habits, owing to its extreme timidity and watchfulness, plunging at once through its hole in the ice, on the approach of danger. Acerbi says, that "neither their teeth nor paws have any share in making these holes, but that it is performed solely by the breath." These openings are kept clear, but the surface is allowed to freeze over partially, so as to conceal them effectually, except from very experienced eyes.

THE HARP SEAL (*Phoca Groenlandica*) is met with principally on the coasts of Labrador; it is remarkable for changing its color annually till it attains a mature age, being in the first year cream-colored, in the second gray, in the third and fourth spotted, and in the fifth it has a black mark on its back like two crescents with their horns uniformly directed towards each other.

THE FŒTID SEAL (*P. fœtida*) frequents the fixed ice near the land, never relinquishing its haunts when old. The adults are remarkably fœtid, the odor even tainting their flesh. It is found occasionally in Labrador, but its chief haunts are the shores of Greenland.

THE URSINE SEAL (*P. ursina*) is a large animal, averaging eight feet when grown to maturity. This species, though gregarious, keep in separate families, each male having his seraglio of from eight to thirty females, over which he watches with incessant jealousy; it is very fierce. The only part of America in which this is found is the group of islands between this Continent and Kamschatka; and, like the sea otter, is only seen between the 50th and 60th parallels. They are there migratory, arriving at those islands in June, and remaining till September. They differ from all other seals in having the anterior limbs entirely at liberty, and not at all enveloped by any integument.

THE MORSE or WALRUS (*Trichechus Rosmarus*) is closely allied to the seal in appearance, but chiefly distinguished from it in size, (weighing from 1,500 to 2,000 lbs.) It has also two large canine teeth or tusks, measuring from ten to twenty inches, directed downwards from the upper jaw, and curved towards the body. These are probably made use of by the animal as an assistance in climbing the ice, acting on the principle of a hook, as well as in self-defence. Scoresby relates an instance in which two Walruses, when attacked,  attempted to destroy the boat containing their assailants, by rising alongside and hooking their tusks over its sides, evidently with the purpose of swamping it. Cuvier, in speaking of its teeth says, " that it has a system altogether peculiar, as it is not better adapted for bruising vegetable, than for cutting animal food; the teeth from their structure, must act like a pestle and mortar." When undisturbed they are fearless and inoffensive, slow and clumsy on land, but easy of motion in the water. They are found abundantly on the coasts of Davis' Straits, and the Magdalen islands; individuals have been met with on Anticosti, which, as well as one or two that are stated to have been seen on the shores of the Lower St. Lawrence, probably were unwilling visitors on some drifting iceberg. Their peculiar habitat is the extreme northern coast of this continent.

CHAPTER VIII.

MARSUPIALIA — THE OPOSSUM — DESCRIPTION AND HAUNTS. — RO-
DENTIA — THE BEAVER, ITS HABITS, AND DESCRIPTION OF ITS
HOUSE AND DAM. — THE MUSQUASH OR MUSKRAT AND ITS
PECULIARITIES.

The Marsupialia or Pouched Animals, who get their name from the Latin word *marsupium*, a bag, were entirely unknown to the ancients, the Opossum being peculiar to America, and the Kangaroo to New Holland.

Description.—About the size of the common cat; head like a fox: ears large and naked: mouth deeply cut, opening wide: tail long and tapering, hairy towards the body, the remaining part covered with scales, prehensile: legs short, color gray or mixture of black and white.

The females of this species have underneath, a pouch formed by an elongation or fold of the skin of the belly, supported by two long bones connected with the muscles of the belly, and articulated or jointed at the pubis. This peculiar construction first occasioned their original describers to be considered rather as inventors than trustworthy witnesses, and it was a considerable time before they were correctly represented. Buffon, though learnedly and elaborately exposing the errors of other writers with respect to this singular animal or class of animals, has himself given a very inaccurate description of it, confounding the opossum of Virginia with the Australian Kangaroo, but giving for the former, a figure unlike either, though between both. According to Demarest, they differ from all other animals in the production of their young, which are apparently brought forth *premature ;* for when first discovered in the external pouch, they are incapable of movement, exhibiting but slight traces of limbs, and are attached to the teats of the mother, of which they are unable to resume their hold if it be broken, these teats lying, from ten to twelve in number, within the pouch. They remain thus attached until strong enough to move about, but they continue to take refuge in the pouch till they have attained the size of a rat. Godman objects to this term *premature*, saying that their birth is perfectly mature and regular, though apparently premature when compared with other animals.

There are said to be twenty-one species (probably varieties) of opossum on the North American Continent, of which the Virginian (*Didelphis Virginiana*) is the commonest, being met with from the Canadian lakes to Paraguay. In its wild state, this animal scopes out for itself a burrow near the bushes in the neighborhood of habitations. He is essentially a nocturnal animal, sleeping by day, and prowling about during the night, living on such small birds and quadrupeds as he can catch. He mounts the trees, penetrates into the poultry yards, attacks the hens, sucking their blood, devours their eggs, and when he is satisfied returns to conceal himself at the bottom of his burrow. When he cannot obtain flesh, he eats fruits and other vegetables, and occasionally reptiles and insects. He is a capital climber, for which his sharp claws and prehensile tail are well adapted, sometimes suspending himself by that appendage from the branch of a tree, on the watch for some luckless bird or squirrel, that may come within his reach; he also leaps like a squirrel from tree to tree with great agility. With habits of life analogous to the polecat and the fox, he is much less cruel and sanguinary, nor is he so well furnished as they are with the means of defence.

In confinement he is tame and amiable, but uncleanly and disagreeable. When attacked in the woods, and finding no other means of escape he rolls himself into a ball, and if on a tree, will fall to the ground and pretend to be dead, though unhurt and ready to run away the instant it can be done safely; and it is said that no infliction of pain will make him shew the least signs of life. This is what is called " playing possum" in the West, a saying which has become almost universal.

The Opossums are very tenacious of life, a fact which has led to a Virginian proverb, " If a cat has nine lives, an opossum has nineteen." They treat their young with the greatest affection, receiving them into their wonted place of protection on the least alarm, and when there is not time for this, the little ones wind their tails around that of the mother, and thus all escape. They do not drink by lapping, but by suction. There is not much probability of the species becoming extinct, as the young amount to from twelve to sixteen at a birth, and their nocturnal habits do not require them to remove far from the haunts of men.

The *Glires* or *Rodentia* constitute the next family, so called from the two large cutting teeth found in each jaw. They have no canine teeth; live almost exclusively on vegetable food; have two incisors in each jaw, large, strong, and remote from the grinders. This order comprises a great number of the smaller quadrupeds. Of these, THE BEAVER (*Castor Fiber*,) is the most useful to man.

Description.—Fur dense, body thick and clumsy, head broad and conical, eyes small and black, ears short and rounded, and almost concealed in the fur; tail broad, flattened, oval, naked, and scaly; back arched.

The Beaver is an animal possessed of a large amount of instinct; but though Buffon and many early writers have made it appear endowed with extraordinary sagacity, a rational view shows us, that it is not more blessed than any animal in its own sphere. The greater part of its history as generally accepted and related is *fabulous;* and it is only from the most authentic sources that we get a reliable account of its habits. The construction of dams by them is so well known as to need no comment—memorials of them remaining in every state, though the animal itself has ceased to exist in them. According to the Indian accounts, the beaver builds the north side of his house the thickest, the more effectually to resist the cold. It feeds on the bark of trees, shell fish, and the roots of the pond lily (*nuphar luteum*). Between the skin and the roots of this animal's tail, lie two oval glands, containing a pure strong oil of a rancid smell; this was originally the castor oil of the early medicos, and had need to be a costly drug. We all doubtless remember the story of the beaver when pursued biting out those glands and casting them before the hunter, alluded to by Juvenal, and handled by Æsop in his fables. But it may be some satisfaction to the consumer to know that the present drug, although analagous in name, is prepared from a bean, and is only allied to the animal oil in name. There is only one known species of this animal. Its principal habitat now is the Hudson Bay Territory; and its skin is mostly relied upon by the Indians as their means of barter with the white man.

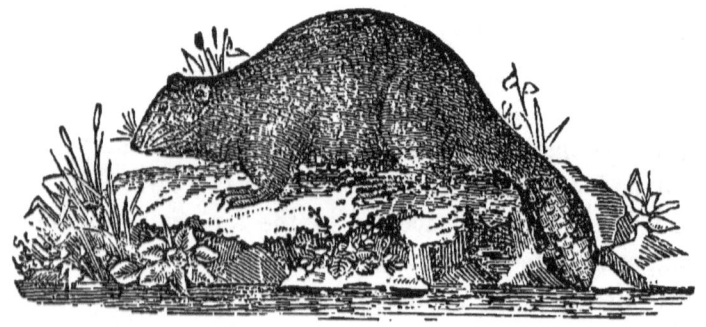

As it has been adopted together with the maple leaf as a Canadian emblem, it is deemed advisable to append the following account by Hearne, who studied the habits of this animal for twenty years, in the Hudson Bay territory, and which is pronounced by Dr. Richardson, who himself had the best opportunity for ascertaining its truth, to be the most correct and free from exaggeration, which has ever been published.

"Where beavers are numerous they construct their habitations upon the banks of lakes, ponds, rivers and small streams; but when they are at liberty to choose, they always select places where there is sufficient current to facilitate the transportation of food and other necessaries to their dwellings, and where the water is so deep as not to be frozen to the bottom during the winter.

The beavers that build their houses in small rivers and creeks, in which the water is liable to be drained off, when the back supplies are dried up by the frost, are wonderfully taught by instinct to provide against that evil, by making a dam quite across the stream, at a convenient distance from their houses. The beaver dams differ in shape according to the nature of the place in which they are built. If the water in the stream have but little motion, the dam is almost straight; but when the current is more rapid, it is always made with a considerable curve, convex toward the stream. The materials made use of are drift wood, green willows, birch and poplars, if they can be got; also mud and stones, intermixed in such a manner as must materially contribute to their strength. In places which have been long frequented by beavers undisturbed, their dams by frequent repairing, become a solid bank, capable of resisting a great force both of water and of ice. The beaver houses are built of the same materials as the dams, and are always proportioned in size to the number of inhabitants. Instead of order and regulation being observed in rearing their houses, they are of much ruder structure than their dams; for it has never been observed that they aim at any other convenience in their houses, than to have a dry place to lie on. It frequently happens that some of the larger houses are found to have one or more partitions, if they deserve that appellation; but it is no more than a part of the main building, left by the sagacity

64   ANIMALS OF NORTH AMERICA.

of the beaver to support the roof. The entrance is always under water: and travellers who assert that beavers have two doors to their houses, one on the land side, and the other in the water, seem to be less acquainted with these animals than those who assign them elegant apartments. Such a

construction would render their houses of no use, either to guard them from the attacks of their enemies, or against extremely cold weather.

All their work is performed in the night; and they are so expeditious that in the course of one night I have known them to have collected as much mud as amounted to some thousands of their little handsful. They cover the outside of their houses every fall with fresh mud, and as late in the season as possible, which, freezing as hard as a stone, protects them from their common enemy the wolverine; and as they frequently walk over their work, and sometimes give a flap with their tail, particularly when plunging into the water, this without doubt has given rise to the vulgar opinion that they use their tails as a trowel, with which they plaster their work; for the flapping of the tail is only a habit which they always preserve, even when they become tame and domestic, and more particularly so when they are startled.

The animal mostly allied to the beaver is THE MUSKRAT or MUSQUASH (*Fiber Zibethicus*). It probably gets its name from the musky odor it emits, which even is prevalent in the tail long after it is dried. Unlike the beaver, it remains without fear at the advance of civilization, and doubtless owes its security to its nocturnal habits. Its burrows are very injurious to mill-dams and embankments. Bartram says that in Carolina, Georgia, and Florida, it is never met with within a hundred miles of the coast, showing the care of the Creator in not distributing it, where so much depends upon embankments.

Its favorite food is the *calamus*, or sweet scented rush, the root of which it devours with avidity : it also feeds on the fresh-water mussel. Godman denies its being pisciverous, but if he had seen the heaps of mussel shells bearing marks of teeth accumulated here and there round some flat stones just above the water in the lakes of New York state, and sundry other northern waters, he would have recalled his assertion. It is found all through the temperate portions of this continent, and is peculiar to it. According to Dr. Richardson, from four to five hundred thousand of its skins are annually exported to England, principally to be used in the manufacture of hats.

CHAPTER IX.

FIELD MICE—DESCRIPTION—THE MEADOW MOOSE—THE MARSH CAMPAGNOL—THE HAIRY CAMPAGNOL OR COTTON RAT—THE WOOD-RAT—THE LEMMING.—THE BLACK RAT—THE COMMON MOUSE—THE POUCHED RAT AND JUMPING MOUSE—THE WOOD CHUCK.—QUEBEC, FRANKLIN, PARRY'S AND HOOD'S MARMOT.—THE PRAIRIE DOG, AND DESCRIPTION.

FIELD MICE (*Arvicolæ*). There are thirty-four known species, though here we only enumerate those most likely to be met with.

Description.—Color grayish-brown above; yellowish lead color below: eyes moderately large and prominent; opening of the ears large; tail short and sparsely covered with hairs.

THE MEADOW MOUSE (*Arvicola riparius*) is the most common of this species: and at times they become so greatly multiplied as to do much injury to the stacks of hay and grain. They have their burrows in the banks of streams, and under old stumps and logs; and numerous furrows may be seen in places where the little animals are plentiful along the roots of the grass, forming lanes in which they may travel in various directions from their burrows. Their nests are sometimes constructed in their burrows, and are also found at the season of hay harvest in great numbers among the vegetation on the surface of the ground. Were it not for the extraordinary fecundity possessed by these creatures, producing six or eight at least three times a year, they would long ago have become extinct, for the owl, the hawk, the fox, the crow, the cat, &c., all combine to check their undue multiplication.

THE MARSH CAMPAGNOL (*Arvicola Floridanus*) has been partly described by Ord, further than which nothing definite is known of it. He says it is frequently seen at high tide; sitting on floating reeds in the marshes, patiently waiting for the receding of the waters. It is peculiar to the extreme Southern States.

The Hairy Campagnol (*Arvicola Hispidus*) was also discovered by Ord in East Florida, and is known nowhere else; it is sometimes called the cotton rat, from its nest being made entirely of cotton.

The Wood-Rat (*Mus Floridanus*) is found throughout the South-western States. Bartram thus describes its nest: " It builds a conical pillar three or four feet high, of brush and dry woods, so interwoven that it would take a bear or wildcat so long to pull it to pieces as would enable it and its young to escape. It is supposed to have been once common, but to have given place to the black rat." In Mississippi and Florida it takes to the houses of the settlers as eagerly as the common rat. It is not known now in the Atlantic States or in Lower Canada.

The Lemming (*Mus Hudsonicus*) differs from the European lemming only in habits, by not migrating in bodies. This may be accounted for by some peculiarity in the soil; and its stock of food probably never failing here as in Europe, it is not compelled to tend its course southwards in search of sustenance. It is met with in the northern part of the Hudson Bay territory.

Of the Rat tribe, we find everywhere the common pest (*Mus decumanus*), probably so named from his decimating or taking tithe of everything that falls in his way. This rat is not indigenous, but imported from Europe, coming originally from Asia. The graphic character given it by Dr. Godman, will not be disputed by any who are acquainted with it. " It must be confessed," he says, " that this rat is one of the veriest scoundrels in the brute creation, though it is a misfortune in him rather than a fault, since he acts solely in obedience to the impulses of his nature. He is by no means as bad as the scoundrels of a higher order of beings, who endowed with the superior powers of intelligence, still act as if they possessed all the villainous qualities of the rat without being able to offer a similar apology for their conduct. He is one of the most impudent, troublesome, mischievous, wicked wretches that ever infested the habitations of man. When he gains access to a library, he does not hesitate to translate and appropriate to his own use the works of the most learned authors, and is not so readily detected as some of his brother pirates of the human kind, since he does not carry off his prize entire, but cuts it into pieces, before conveying it into his den." The only benefit he affords to man is from his skin; large quantities being exported from California (where they are an article of traffic among the Chinese population), in a salted state to France; whence after undergoing certain operations and manipulations there, these loathsome peltries emerge again into the world in the shape of—*kid gloves!* the finest so called kid being made from rat skin. The elegant white leather in druggists' cases which is so elaborately tied over scent bottles, is also the skin of our detested friend.

THE BLACK RAT (*M. Rattus*) has disappeared so entirely before the previous species as to be almost extinct; the common mouse (*Mus Musculus*), likewise an importation, being probably introduced in bales of merchandise, is abundant every where; it is very prolific, producing several litters in the year; and we have the authority of Aristotle two thousand years ago, that from a pregnant female enclosed in a chest of grain, 120 individuals were created in a few months.

THE POUCHED RAT (*Pseudostoma bursarius*) is found in Florida and the extreme South, and is but little known.

THE JUMPING MOUSE (*Gerbalis Canadensis*) is found from Canada to Pennsylvania. This timid and active little creature is abundantly met with in meadows and grain fields, and when not in motion might easily be mistaken for the common field mouse; but its usual mode of progression is very different. Sometimes running on all fours, it more commonly moves by leaping on its hind legs, particularly when pursued. Its hind legs are more than twice the length of the fore ones; In this respect it resembles the Jerboa of Europe.

Mr. Booth, of Orange County, New York, gives the following description of this almost unknown little animal:—"In cross ploughing some years ago, my attention was taken up by seeing something move off from near my plough-share, over ridges and furrows, bearing some resemblance to an old withered oak leaf. I pursued it, when it proved to be one of the above—a female, with four young ones attached to its teats." He says, also from observation, "they are never seen in clear daylight, unless disturbed."

THE LABRADOR JUMPING MOUSE (*Meriones Americus* or *Mus Leucopus*), found only in Labrador and the Hudson Bay territory, completes this family as far as known; but probably the field for discovery among them is not exhausted.

THE MARMOT tribe is represented in America, first by THE COMMON WOODCHUCK (*Arctomyx Monax*), sometimes called

the Ground-hog, or Maryland Marmot. This animal when feeding, frequently rises on its haunches to reconnoitre, raising its fore feet like hands. In this position, when the weather is fine, it will sometimes sit for hours at the entrance of its hole.

On the approach of cold weather, it closes the passage between itself and the surface of the ground, spending the winter in a torpid state. It is a cleanly animal, easily capable of domestication. When its retreat is cut off, it fights hard, its bite with its long and projecting incisors being very severe; the dog that attacks it, showing by its bloody nose and hanging tail, that he has been severely punished or fairly worsted in the battle.

The Genus Spermophilus is distinguished from the Arctomyx, especially in having check pouches, and being much more active and lively. Of these the Quebec Marmot, L. *Tredecim-lineatus*, and Franklin's Marmot, *S. Franklini*, are both found in the Hudson Bay territory; but little is known respecting them. Parry's Marmot, *S. Parryi*, is peculiar to the Arctic regions: Hood's Marmot, *S. Hoodii*, is said to have on its back as many stripes as are displayed on the star spangled banner; and finally, so far as known, the Prairie Dog, *Spermophilus Ludovicianus*, associated among travellers with the burrowing owl and the rattlesnake. For a full account of this animal we are indebted to Captain Marcy, who says :—

"These gregarious and interesting little animals, called 'prairie-dogs,' or more properly the 'prairie marmots,' are found assembled in communities or villages throughout most of our extreme western prairies, from the Missouri River to the Rio Grande, and have often been described by travellers; but as there are some facts connected with them, which I have never seen mentioned, I will add a few remarks to what has already been said.

"These animals, in selecting a site or position for their towns or warrens, generally choose a very elevated and level spot upon the open prairie, which has induced me to suppose that they do not require water—that element without which most other animals soon perish. I have often seen their towns upon the table lands of New Mexico, at a distance of twenty miles from any water upon the surface of the ground, and where it did not seem probable that it could be obtained by excavation; and as there is seldom any rain or dew upon these elevated *mèsas* during the summer months, and as they do not wander far from their burrows, I think I am warranted in coming to the conclusion that they require no other aqueous sustenance than that which they receive from the short grass which constitutes their food. Their burrows are generally placed about fifteen yards apart, and each constitutes the abode of five or six occupants.

" The towns vary much in magnitude, some only covering the space of a few acres, while others are spread over a surface of many miles. I passed through one upon the head waters of Red River, which was twenty-five miles in length. Supposing it to have been of the same length in other directions, it would cover an area of 625 square miles.

" They appear to delight in sporting with each other about the entrances of their holes, and may always be seen in pleasant weather, frolicking, running, and barking throughout the whole town; but at the slightest sound, or the least approach of danger, they make a precipitate retreat to their burrows, dropping themselves partly in with their heads above the ground, and their eyes intently fixed in the direction of the intruder, at the same time flourishing their tails from side to side, with nervous jerks, and keeping up an incessant barking until the danger approaches too close, when they suddenly disappear beneath the ground, and the town, from ringing with their music, becomes in an instant as silent as the grave.

" That these animals hybernate and pass a portion of the winter in a lethargic or torpid state, is evident from the fact that they do not, like the squirrel, lay up sustenance. When they first feel the approach of the sleeping season, (generally about the last days of November,) they carefully close all the passages to their dormitories to exclude the cold.

"They remain housed until the warm days of spring, when they remove the obstructions from their doors, and again appear above ground as gay and frolicsome as ever. In the early part of winter they are sometimes seen reopening the entrances to their domicils while the weather is still cold and stormy, but mild and pleasant weather is sure to follow, from which it appears that instinct teaches them when to anticipate good or bad weather, and they make their arrangements accordingly.

"It has been said that the rattlesnake, and a species of small owl, are always found burrowing with the prairie-dog in the most perfect harmony. The snake is sometimes seen in the towns, but he is by no means a welcome guest with the proprietors of the establishment, and only resorts there to prey upon the dog. One that was killed by our party was found, upon examination, to have swallowed a full grown marmot. The owl is always seen sitting near the dogs at the entrance of the burrows, but I have never known them to enter or emerge from them. On approaching near they always fly away."

CHAPTER X.

SQUIRRELS AND THEIR HABITS—THE GRAY AND THE FOX SQUIRREL,—THE BLACK AND THE RED—THE GROUND SQUIRREL OR CHIPMUNK—THE FLYING SQUIRREL—THE PORCUPINE — ITS HABITS AND USE OF ITS QUILLS—THE HARE—THE VARYING HARE.

The next family is that of the SQUIRRELS (*Sciuridæ*), nearly all of them living on trees, for which purpose their long flexible toes with acute nails, enable them to leap from tree to tree, rarely missing their hold.

Description.—Their tails are long and bushy; eyes large; body elongated; ears erect: they feed on nuts, seeds, grain, &c.

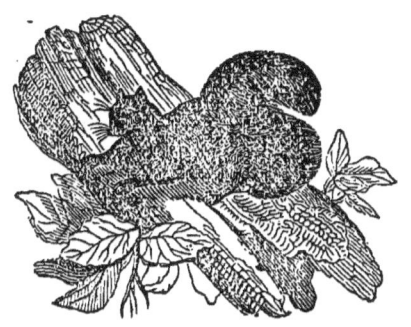

Foremost amongst these lively little animals, stands THE GREY SQUIRREL (*Sciurus leucotis* or *cinereus*), found everywhere through the continent. This, as well as some of the other species, in some years becomes exceedingly multiplied, and then perhaps for some years very few of them will be seen. This sudden increase and diminution of their numbers, seems to depend upon two causes, the supply of food, and the severity of the winters. Their great multiplication generally follows a mild winter, which has been preceded by a productive summer. The Grey Squirrel prefers woods abounding in oak, walnut, butternut or hickory trees, as affording him most food : during the fall he collects a supply for the winter, which he deposits in some hollow tree; often also he accumulates large stocks of nuts in various places, covering them with leaves, and seeming to forget the locality; for the writer remembers in Westchester Co., N. Y., one afternoon in early winter, that five or six boys rambling with him came upon such a number of these heaps of hickory nuts, that they were unable to carry all away. One peculiarity of this species is its singular and distant migration in large bodies. Bachman has furnished an interesting account of an extraordinary migration of this sort, which he witnessed near Albany, N. Y. On that occasion troops of squirrels suddenly and unexpectedly made their appearance; they swam the Hudson River with their bodies and tails wholly submerged; many were drowned, and those that reached the opposite bank were so wet and fatigued, that they were easily killed. An unusual and general failure of their food is supposed to be the motive for such migrations. The popular belief that the males emasculate each other is without foundation.

The Fox Squirrel (*S. vulpinus*) is probably only a large species of the preceding, as there are no peculiarities to mark a difference, except size and robustness. It is confined to the Southern States.

The Black Squirrel (*S. Niger*) is very similar in habits, &c., to the gray squirrel, but seldom found south of Pennsylvania; it is said to disappear before the grey.

The Red Squirrel (*S. Hudsonicus*) extends abundantly from the Arctic Circle to Tennessee; he is a noisy little fellow, with a twittering note of *chick-a-ree*, which has suggested that, as one of his popular names. He is to game what the magpie or jay are in England,—a most watchful tattling spy; for however stealthily the sportsman may move, or in ambush await his

game, this, nature's constable, on first sight sets up his shrill cry, putting all else around on their guard. His favorite haunt is a cedar swamp. His habits are materially influenced by climate; at the north he forms deep burrows in the earth, under the roots of trees to protect himself from the cold, whilst further south he contents himself with the hollow of a tree.

THE STRIPED SQUIRREL (S. *striatus*), better known under the name of Ground Squirrel, Hackee, or Chipmunk (the latter probably its aboriginal name), is met with almost all over the continent. It differs from the preceding species in having its chief residence in the ground, while the others take to trees. It is usually seen running along fences, and stone walls which afford it a ready retreat. Under these it usually makes its burrow, and lays up its store for winter use. A favorite spot is the centre of some decayed or decaying stump. It seldom ascends trees; is of an irritable nature, resisting every attempt at domestication. When alarmed, it immediately takes to its hole, which it enters with a peculiar shrill cry indicative of safety, as much as to say "catch me now if you can." In the fall of the year it is very busy conveying grain, &c., to its winter quarters, instinct warning it of the approaching season; and really if some of our own species would take this provident little fellow as an example, it would relieve charitable societies of many of their cares! There are many other species of Squirrel, but locally distributed.

The next species is the *Pteromys*, or Flying Squirrel.

Description.—Skin dilated on the sides, from the fore to the hind legs, forming a sort of parachute: from the fore arm is a bony appendage supporting the membrane. By the aid of this membrane it darts from tree to tree, not by a movement of it as in the bats, but by sailing downwards obliquely, and rising suddenly when within a few inches of the tree it wishes to alight upon.

THE FLYING SQUIRREL (*Pteromys volucella*) is met with in

all parts of the Continent, though never so greatly multiplied as some of the preceding species. It usually inhabits the hollows in trees, is seldom seen except in the twilight, and in its sailing movement is aided by the broadly expanded tail, which doubtless acts also partially as a rudder. It seems to ascend some high tree, throw itself off, alighting on another near the ground, ascend that again in turn, and proceed in like manner to another, thus progressing a considerable distance without touching the ground. It is gentle in its disposition,

easily domesticated, fond of warmth. At twilight it arouses itself, and affords much entertainment by sailing about the room, always commencing its flight by climbing a table, chair, or on to a shelf from whence it may start its voyage. When it sleeps, it rolls itself up, and wraps its large flat tail over its head and limbs, so as completely to conceal them, and give it the appearance of a simple ball of fur. There are two other species of it known, the *P. Sabinus*, like the foregoing in every respect but size, in which it excels, and *P. Oregonensis*, with ears longer and more open.

THE COMMON PORCUPINE (*Hystrix* or *Erethizon dorsata*) though formerly very common, is now confined to the wooded and mountainous districts.

Description.—Hair rather long, thick, and interspersed with spines or quills, varying from one to four inches in length; these quills are black at the tip, brown below, white at the base. Tail not prehensile, thick.

It is remarkable on account of its quills or spines, intermingled in the hair, on nearly all parts of its body; and as it runs very badly, and is moderate and awkward in all its movements, it relies mostly on its quills for defence and safety. When its enemy approaches, if allowed sufficient time, it will generally retreat to some fissure in the rocks, or take refuge in the top of a tree, which it ascends with facility; but if overtaken it places its head between its fore legs, draws its body into a globular form, and erects its spines projecting in all directions. The old theory of the porcupine darting his quills has exploded long ago. These quills are eagerly sought after and highly prized by the aborigines on all parts of the continent, and are used by them in various ways as ornaments of their dresses, pipes, and war instruments. For this purpose they are dyed of several rich and permanent colours, cut into short pieces, strung upon threads or sinews, and then wrought into various forms and figures upon their belts, buffalo robes, mocassins, &c., and in these adornments they show considerable ingenuity and a great deal of patient perseverance.

The porcupine is a sluggish, solitary animal, seldom venturing far from his retreat in the rocks. His food consists of fruits of different kinds, roots, herbs, and the bark and buds of trees. The flesh is esteemed by the Indians the greatest luxury. In confinement it will eat bread and vegetables from the hand, come when called, and exhibit fondness for its owner.

THE HARE (*Lepus*) next engrosses our attention.

Description.—Upper incisors placed in pairs; head large; ears long; eyes large, projecting laterally; hind feet long; bottoms of feet, hairy; tail, very short and elevated. Never burrowing like the European rabbit.

THE AMERICAN RABBIT (*Lepus nanus* or *Sylvaticus*) though, strictly speaking, a hare, is the most common of the species throughout the continent. Fleming says, that the American hare and European rabbit so nearly resemble each other, that it puzzles zoologists to assign distinguishing marks. The only points in which they differ are as follows: The nest of the hare is open, while that of the rabbit is closed, and lined with its own fur. The young of the hare are brisk, have eyes and ears perfect, their legs in a condition for running, and their bodies covered with fur at birth. The young of the rabbit have their eyes and ears closed, are unable to travel, and are naked. It has been said that the American rabbit forms burrows, but this is decidedly a mistake, probably arising from its occasionally taking refuge when hard-pressed in the holes of foxes or wood-chucks.

THE VARYING HARE (*Lepus Americanus*) sometimes called the Northern hare, is found all through the mountainous regions of the West, and is not dissimilar to the Alpine hare of Europe. In the winter season it becomes perfectly white; it is less prolific than the preceding, hence its greater scarceness. Warden says this hare differs from the Alpine, by changing from *gray-brown* to white; its ears are shorter; its legs more slender. According to Godman, scarcely sufficient distinction exists to warrant the idea of its being other than the Alpine; and in this the writer agrees, pronouncing it only a variety. In summer the hares feed on grass, juicy herbs, and the leaves and buds of shrubs, but in winter, when the snow is deep they gain a precarious subsistence from the buds and bark of the willow, the birch, and the poplar. When pursued, the rabbit (so called) soon becomes wearied, but the hare is so fleet that being in no fear of being overtaken by its pursuers, it seeks no concealment; it has been known by measurement to clear 21 feet at one bound, and its body is so light in comparison with its broad, furry feet, that it is enabled to skim easily along the surface of deep snows, while the wearied hounds plunge in at every bound. There are thirteen other species, but all confined to Arctic latitudes.

CHAPTER XI.

RUMINANTIA.—DESCRIPTION OF THE PROCESS OF CHEWING THE CUD—USEFULNESS OF THE ORDER—THE DEER FAMILY—THE ELK OR WAPITI—HABITS AND DESCRIPTION—EXTRACT FROM LONG'S EXPEDITION.

The order Ruminantia, Pecora, or Cud-chewers, next follows, peculiarly distinguished by having no incisive teeth in the upper jaw; their feet are all two-toed, covered with two hoofs, having the appearance of a single one, cleft in the middle. They are altogether herbiverous, and have the power of returning the food to the mouth after it is once swallowed, for a second mastication; and as the process of chewing the cud is not generally known, we may as well give an explanation of it here. These animals are possessed of four stomachs, the first called the *rumea* or paunch, being of such capacity as to receive the large bulk of vegetable matter coarsely bruised by the first mastication. Passing into the second stomach, the *reticulum* or honeycomb (so called from having a beautiful internal membrane of polygonal acute angled cells), the food is here moistened and formed into little pellets, which are then thrown up into the mouth to be again chewed. It is then swallowed the second time in a fine pulpy state, and passes into the third, the *omasum*, and finally into the fourth, the *abomasum*, or reed, which is of a pear-shape, and wrinkled, corresponding to the human stomach. Here it is digested by the action of the gastric juice, and its nourishing parts absorbed and thrown into the circulation for the growth and renovation of the living system. This gastric or stomach juice, is a colorless liquid, secreted or prepared by the stomach, and by which the process of digestion is carried on.

The first stomach, in which the food is received, is very large when compared with the others. This is a kind of storehouse or receptacle where the food by warmth and moisture is prepared for the second chewing. When this sack or pouch is well filled, the animal generally retires to a shady place and lies down, when the process of ruminating begins. The alternate motion of the cud, as it pasess up and down, is quite apparent in many ruminants.

To this order man is most indebted, his food being obtained from their flesh, and their hides, horns, bones, hair, milk, and even blood being hourly in demand: yielding valuable service as beasts of burden; feeding on the verdure of the land, which by converting into their own flesh they prepare for the use of man, nine-tenths of whom never think of looking on them with an eye of admiration or wonder, or of enquiring into the habits or the formation of this useful family of animals.

The general structure of this order is strikingly well adapted to their habits and wants. In general, their legs are long in proportion to the length of their bodies, and their backbone is not only of great length but highly flexible, both of which are conditions favorable to great activity and freedom of motion. Their ears are generally long and movable in all directions, so that sounds may be caught from whatever quarter they come. This is a provision of great consequence, since, while feeding, their ears are turned in a direction contrary to their sight, and hence they may be warned of danger from before as well as behind. Their eyes are situated at the sides of the head, and thus in addition to the usual range of vision of other animals, they can see behind as well as before them. It will be obvious that these are designed and merciful provisions, since these animals when pursued, can hear the direction of their followers, and see them also without stopping, while effecting their escape at full speed.

Their means of defence are their horns and hoofs, in

the uses of which some of the members are exceedingly expert and effective : in the deer family, including the elk and moose, the fore-feet are generally the most destructive weapon, the largest dogs being sometimes killed by a single blow, and there are instances of several hunters in California having met with a similar fate from the stamping of the Elk.

Great confusion has arisen under this class, relative to the species of North American Deer, partly from imperfect observation, and partly from varieties dependent on age and sex. In all the species, with the exception of the Reindeer, the male only is crowned with horns or antlers. These appendages are shed, or fall off, every year, but are renewed with increased size, as the animal advances in age. The peculiar mode in which the antlers are developed, and in which the separation is effected between them and the head, forms one of the most curious phenomena in the animal economy.

The ELK, or WAPITI (*Cervus Canadensis*) has been commonly confounded with the Moose, and with the common Stag of Europe, and has passed under various authors' hands by the name of Stag, Red Deer, Gray Moose, Wapiti, Round Horn, Cariboo, &c. It is only very recently that it has been distinguished as a separate animal; and the confusion attending this varied nomenclature, has been rather increased than diminished by those who have attempted its removal by reconciling the discrepancies of books, instead of appealing to the proper and infallible authority—nature.

Description.—The size and appearance of the Elk are imposing ; his air denotes confidence of great strength ; while his towering horns exhibit weapons capable of doing much injury when offensively employed. The head is beautifully formed, tapering to a narrow point; the ears are large and rapidly movable ; the eyes are full and dark ; the horns rise loftily from the front, with numerous sharp-pointed branches, which are curved forwards; and the head is sustained upon a neck at once slender, vigorous, and graceful. The beauty of the male Elk is still farther heightened by the long forward, curling hair, which forms a sort of ruff or beard, extending from the head towards the breast, where it grows short, and is but little different from the common covering. The body of the Elk, though large, is finely proportioned; the limbs are small and apparently delicate, but are strong, sinewy, and agile. The hair is of a bluish gray color in autumn; during winter it continues of a dark gray; and at the approach of spring it assumes a reddish or bright brown color, which is permanent throughout summer.

The Elk has at one time ranged over the greater part, if not the whole of this continent. It is still met with in the remote and thinly settled parts of Pennsylvania, but the number is small. It is only in the Western wilds that they are met with in considerable herds. They are fond of the green forests, where a luxuriant vegetation affords them an abundant supply of buds and tender twigs; or of the great plains, where the solitude is seldom interrupted, and all-bounteous nature spreads an immense field for their support. The Elk sheds its horns about the end of February, or beginning of March; and such is the rapidity with which the new horns shoot forth, that in less than a month they are a foot in length. The whole surface of the horn is covered by a soft, hairy membrane, which, from its resemblance to that substance, is called velvet; and the horns are said to be "in the velvet" until the month of August, by which time they have attained their full size. After they are fully formed, the membrane becomes entirely detached; and this separation is hastened by the animal, who appears to suffer some irritation, which causes him to rub them against the trees. The velvet, in its hanging state, closely resembles a tangled mass of *cobwebs*.

The Elk is shy and retiring; having very acute senses, it receives early warning of the approach of any human intruder. As soon as the hunter is fairly discovered, it bounds along for a few paces, stops, turns half round, and scans its pursuer with a steady gaze, then throwing back its lofty horns upon its neck, and projecting its taper nose forward, it springs from the ground, and advances with a velocity which soon leaves the object of its dread far out of sight. But during August and September, when the horns are in perfect order, it employs them and its hoofs with great effect, and the lives of men and dogs are endangered by coming within its reach. When at bay, it fights with great eagerness, as if resolved to be avenged. The following incident, from Long's Expedition to the Rocky Mountains, will, in some degree, illustrate this statement:

"A herd of twenty or thirty Elk were seen at no great distance from the party, standing in the water, or lying upon the sand beach. One of the finest bucks was singled out by a hunter, who fired upon him; whereupon the whole herd plunged into the thicket, and disappeared. Relying upon the skill of the hunter, and confident that his shot was fatal, several of the party dismounted and pursued the Elk into the woods, where the wounded buck was soon overtaken. Finding his pursuers close upon him, the Elk turned furious upon the foremost, who only saved himself by springing into a thicket, which was impassable to the Elk, whose enormous antlers becoming so entangled in the vines, as to be covered to their tips, he was held fast and blindfolded, and was despatched by repeated bullets and stabs."

CHAPTER XII.

THE MOOSE—DESCRIPTION AND HAUNTS—MODE OF HUNTING IT—THE CARIBOO AND DESCRIPTION—HUNTING ENCAMPMENT—THE COMMON DEER—HABITS AND VARIOUS STORIES RESPECTING IT.

THE MOOSE (*Cerius alres*), the largest of all the species, is the only deer whose appearance may be called ugly. Its name is probably a corruption of the word *Musu*, a term given to it by the Algonquins.

Description.—Its large head terminates in a square muzzle, having the nostrils curiously slouched over the sides of the mouth; the neck, from which rises a short thick mane, is not longer than the head, but is rendered still more cumbrous and unwieldy by wide palmated horns; under the throat is found an excrescence, from which grows a tuft of long hair; the body, which is short and thick, is mounted upon tall legs. These singularities of structure, however, have direct or indirect reference to peculiarities of use adapted to circumstances.

The Moose inhabits the northern parts of both continents, and is in Europe called "the Elk." On this continent it has been found as far north as the country has been explored. Its southern range at former periods extended to the shores of the Great Lakes. At present it is not heard of south of the State of Maine, and is becoming rare also there. In Nova Scotia, Cape Breton, and throughout the Hudson Bay possessions it is found in considerable numbers. The dense forests, and closely shaded swamps of these regions, are the favorite resorts of this animal, as there the most abundant supply of food is to be obtained, with the least inconvenience. The length of limb and shortness of neck, which in an open pasture appear so disadvantageous, are here of the greatest use in enabling it to crop the buds and young twigs of the birch, maple, or poplar; or enabling it to browse on aquatic

plants, inaccessible to other animals. In the summer, it frequents swampy or low grounds, near the margins of lakes and rivers, through which it delights to swim, as this frees it from the annoyance of insects. At this season it regularly visits the same place to drink, of which circumstance the Indian hunter takes advantage, by lying in ambush.

In Nova Scotia and New Brunswick, it is generally hunted in the month of March, when the snow is deep, and sufficiently crusted with ice to bear the weight of a dog, but not that of a Moose. Five or six men provided with knapsacks, containing food for as many days, and all necessary implements for "camping" out at night, set out in search of their game. Having found their animal, they wait till daybreak, when the dogs are laid on, and the hunters wearing large snow shoes follow as closely as possible. The deer does not run far, before the crust on the snow through which he breaks at every step, cuts his legs so severely that the poor animal stands at bay, and endeavors to defend himself by striking with his fore feet, but the arrival of the hunter soon ends his career.

The skin of the Moose is of great value to the Indian, as it is used for tent covers, clothing, &c. It is feared from the rapid destruction of these animals, and the way in which they have diminished of late years, that the species will eventually become extinct.

"In the winter of 1842, twenty three officers," as we are informed by Porter, "of the Grenadier and Coldstream Guards, then in garrison at Quebec and Montreal, killed during a short hunting tour, ninety-three Moose. None of the parties were absent more than fourteen days." But a more remarkable fact, as related by "Frank Forrester," was "the killing of three moose with a common fowling piece, by an officer not reputed to be very crack as a shot, on the Mountain, within a few miles of Montreal, during a morning's walk from that populous city." He also cites another instance of a friend killing seven of these glorious animals on the River St. Maurice, in the rear of the pretty village of Three Rivers, all of which he ran into upon snow shoes, after a chase of about three days.

THE REINDEER, or CARIBOO (*Cervus tarandus* or *Tarandus rangifer*).

Description.—Body robust, and low on the legs; snout thin, with oblique nostrils; ears large; horns usually slender, the main stem directed backwards, terminating in a broad palmated expansion; hoofs rounded; color varies with age; a smooth coat of grayish brown,—beneath the throat and belly, white.

In usefulness, this animal exceeds all others of the northern zone; and it is a curious fact that, though domesticated in Europe and Asia,—where, as beasts of burden, milkers, furnishing food, in the application of their sinews to bow-strings, and by their powers of endurance, they point out the admirable wisdom of the Deity in placing them where the natives have so little to depend on,—yet the North American Indians have never in any way made use of their living services.

It is as yet an unsettled fact as to the identity of the Cariboo or Reindeer of North America with the celebrated beast of draught, and much less is known of it than of the Moose. De Kay, in his history of the State of New York, states it to be much of the same size as the common deer; but it is since ascertained that the adult males are often found fourteen to fifteen hands high. It is this difference of size which has led to the belief that the cariboo is a distinct variety from that which is the chief article of food to the Esquimaux of the western, and domesticated by the Laplanders of the eastern continent. That animal is scarcely found south of the Arctic circle, while the Cariboo is found *here* everywhere north of the 45th and 46th degrees north latitude.

The mode of hunting Cariboo differs in nothing from that of the moose, with this exception, that owing to the inferior weight of the animal, and the pliability of its pastern joint, which bends so completely, at every stride, under him, as to afford a very considerable fulcrum and support in the deep snow, he is able to travel so much longer and so much more fleetly, even through the worst crusts, that it is considered useless to attempt to run him down, when once alarmed and in motion. He must therefore either be stalked silently from the leeward, or shot down at once.

The following description of a hunting party's encampment is so spirit-stirring, that we cannot pass it by: " The first thing to be done on a tramp after Cariboo, is to encamp for the first night, since it is rare that a single day's march carries the sportsman to the scene of action. The arms are stacked, or hung from the branches of the giant pines around the camp; the goods are piled; the snow is scraped away from a large area, and heaped into banks to windward; a tree or two is felled and a huge fire kindled; beds are prepared of the soft and fragrant tips of cedar and hemlock

branches; and the party gathers about the cheerful blaze, while the collops are hissing in the frying-pan, the coffee is simmering in the camp-kettle, and the fish or game—if the

Indians have found time to catch a salmon-trout or two through the ice of some frozen lake, or the sportsmen have brought down a brace or two of ruffed or Canada grouse—is roasting on wooden spits before the fire, with the rich gravy dripping on the biscuits, which are to serve thereafter as platters for the savory broil. Then comes the merry meal, seasoned by the hunters' Spartan sauce—fatigue and hunger; and when the appetites of all are satiated with forest fare, succeed the composing fumes of the hunter's pipe, replenished with 'the Indian weed that briefly burns,' and such yarns as are spun nowhere, unless it be in a forest camp, are told. * * * Awake, while the stars are yet bright and the air keen and cold, the brook, which last night tempered the goblets, this morning laves the brow and replenishes the kettles, and a brief early breakfast precedes the quick tramp through the morning's gloaming. It is a sport for men, not to be essayed by babes or sucklings. No particular fitness is required except stout thews and sinews—to be long-winded and accustomed to field exercise—and, *en passant*, no man roughs it better than a thorough-bred English gentleman; it is the Cockney who first gives himself airs, and everybody else trouble, and then gives—*out!*"

THE COMMON DEER of North America (*Cervus Virginianus*) differs entirely from all the European or Indian varieties of this order. It is smaller in size than the red deer—hart and hind of the British Isles and the European continent—and is far inferior to it in stateliness of character, in bearing, and in the size and extent of its antlers. From the fallow deer of Europe it differs in being much larger, and having branched instead of palmated horns. It is so much larger than the roebuck, and differs from it so greatly in all respects, that it is needless to enter minutely into the difference.

This beautiful animal abounded formerly in every part of this continent, from the extreme northeast to Mexico, or still farther south, and it is even now found in consider-

able numbers wherever the destruction of the forests, and the wanton rapacity of man, have not caused its extinction. In the State of Maine and in Canada it abounds in the great evergreen forests, its worst enemy there being the wolf, as there is perhaps less of the sporting ardor to be found in that section. The loggers and lumbermen there, may occasionally filch time to hunt, by torchlight, a deer or two, or get up a hunt for a bet—in which, by the way, everything that flies or runs, from an owl to a skunk, is brought to bag promiscuously—but hunting, proper and scientific, there is little or none.

"To get deer-hunting in anything like perfection," says Frank Forrester, "we must go into Virginia, the Carolinas, Louisiana, and Mississippi, where the gentlemen of the land, not pent up in cities, but dwelling upon their estates, fearlessly galloping through bush, through briar, taking bold leaps at fallen trees and over deep *bayous* in the forest lands, ride as fearlessly and desperately for the first blood as any country English squire."

The Indians say, and it has also been verified by hunters, that the deer has a great aversion to snakes, especially the rattlesnake, and to destroy them it makes a bound into the air, alighting on the snake with all four feet brought together, repeating it till the reptile is destroyed. The stomach of the deer, with its half digested contents, is a very favorite dish with the Indians, especially when they feed on mosses and buds, and even Europeans have not objected to it. Captain Lyon says he found it to "resemble a salad of sorrel and radishes;" and Hearne says it possesses such an agreeable taste that were it not for prejudice, it would be considered a dainty.

De Kay says it has often been a wonder that while so many horns are cast annually, so few are ever found. This is to be explained by the fact that as soon as shed they are eaten up by the smaller gnawing animals. He has repeatedly found them half-gnawed up by the various kinds of field-mice so numerous in our forests. From the number of its skins brought to market, and calculating the deer destroyed since the settlement of the country, an imperfect notion may be formed of the aggregate numbers and productiveness of its species.— De Kay has made a strange blunder in his Natural History of New York, in saying " it does not appear to extend into Canada;" for it is most plentiful in both the Upper and Lower Province.

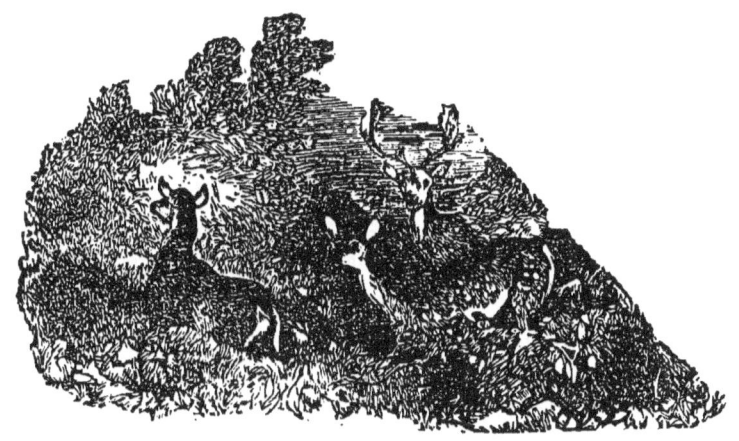

CHAPTER XIII.

THE BLACK TAILED DEER—THE PRONG-HORN OR AMERICAN ANTELOPE—ROCKY MOUNTAIN GOAT—ARGALI OR ROCKY MOUNTAIN SHEEP—DESCRIPTION AND HABITS.

The BLACK-TAILED DEER (*Cervus Macrotis*) is found only near the Rocky Mountains and on the plains of Missouri, preferring generally the prairies to the woods. It was first noticed by Lewis and Clarke in their explorations; they inform us that it resembles its kindred species, except that it does not run at full speed, but bounds along, raising every foot from the ground at the same time. In this it may be compared to the spring-bok of Africa. Its ears are very long, extending half the length of the antler.

The AMERICAN ANTELOPE, or PRONG-HORN (*Antelocapra Americana*), fleeter than the swiftest horse, roams through the Rocky Mountains, keeping entirely to the open ground, migrating in winter to Mexico and California. So swift is it in its movements, and so acute is it in its sense of smell, that man rarely approaches it; and the celerity with which the ground is passed over by it, resembles more closely the flight of a bird, than the motion of a quadruped.

The only reliable accounts we have of it are from Long's Expedition, and Lewis and Clarke's observations, if we except the following from Wilke's exploring expedition: "An antelope was killed in Southern Oregon, near Rogue's River; it was of a dun and white colour, and its hair remarkably soft. The Indians take this animal by exciting its curiosity; for this purpose they conceal themselves in a bush, near its hunting grounds, and making a rustling noise, soon attract its attention, when it is led to advance towards the place of conceal-

ment, until the arrow pierces it. If there are others in company, they will frequently remain with the wounded, until they are all in like manner destroyed."

THE ROCKY MOUNTAIN GOAT (*Capra Montana*) of which very little is known, has for its northern limits the River of the Mountains, and has been met with as low down as 45 degrees north. It is more numerous on the western than on the eastern slope of the Rocky Mountains, but is rarely seen at any distance from them, frequenting the peaks and ridges in summer, and occupying the valleys in winter.

Description.—Nearly the size of a common sheep, with a shaggy appearance, in consequence of the protrusion of the long hair beyond the wool, which is white and soft. The horns and hoofs are black; the horns slightly curved backwards, and projecting but little beyond the wool. The fleece of this goat is said to equal that of the celebrated shawl goat of Cashmere, both in fineness and value, though the skin is spongy and only used for mocassins. Little is known of its haunts; of its habits still less. Were it not for the fleecy nature of its covering, and the shortness of its horns, some analogy might be fancied to exist between this goat and that described by Homer ages ago, utterly lost sight of during all intervening time, and only lately re-discovered in the islands of the Levant. This animal is often confounded with the next species.

THE ROCKY MOUNTAIN SHEEP, or ARGALI (*Ovis Montana*), called also the Cimaron.

Description.—Larger than the common sheep; the ears pointed; the horns which are transversely wrinkled, large, and triangular, are twisted laterally into a spiral; the limbs are slender, and covered with uniform short hair.

These animals are confined exclusively to the Rocky Mountains: they are met with in herds of from twenty to thirty, but are very wary. They feed on the tops of the ridges, with posted sentinels ever watchful; and their great quickness of sight and hearing, render them perhaps the most difficult to approach of all the four-footed game of America. They have immense horns, especially the old males, in whom they are so enormous that from their curving forward and downward to such an extent, they preclude them from feeding on level ground.

It is necessary therefore for them to seek the pasture on steep places above them, or to browse on the long herbage on the margin of the water-courses. They are usually found about grassy knolls skirted by craggy rocks, to which they can retreat when pursued by dogs or wolves. The Indian appellation for them is "the foolish bear," for in the retired parts of the mountains, where no fire-arms have been used, they are quite tame, exhibiting the simplicity of the domestic sheep; but when they have been often fired at, they assume the wild and vigilant character under which they are generally described. In its facility of leaping from crag to crag it resembles the chamois of the Alps.

An attempt was made some years ago to obtain some young of this species and domesticate them in the Scotch mountains for the sake of their fleece, which far excels that of common sheep, but the attempt utterly failed. We cannot do better than close the account of this animal with the following observations from the pen of Frank Forrester, who is about the only writer that has handled the several animals peculiar to the Rocky Mountains. He says:—

"I conceive that this animal is rarely an object of particular systematic pursuit; and that when killed at all, it is almost by accident, during the winter season. While among the herbless crags and awful precipices of those dread mountain solitudes, it is not easy to see it; and when seen, to outclimb and circumvent it, must require that the hunter should be every inch a man. If possible, stalk it having the sun on your back, and in his eyes; or, approach it from the upper to the lower ground; for, as it is its nature to keep the upper ground if possible, it consequently keeps the brightest lookout for an enemy's approach from below; but all depends on the direction of the wind, down which it is impossible to approach it."

CHAPTER XIV.

THE BISON OR BUFFALO—DESCRIPTION—INDIAN MODES OF CAPTURE—EARLY DESCRIPTION OF IT—BUFFALO HUNT—ITS HAUNTS AND ENEMIES.

The next Genus is the Ox, of which the first species is the BISON (*Bos Americanus*), better known under the name of BUFFALO.

Description.—Great disproportion between the fore and hind quarters, partly occcasioned by the hump over its shoulders, which diminishing as it extends backwards, gives obliquity to the outline of its back. The horns are shorter than in any other species, nearly straight, exceedingly strong, and planted widely asunder at the base. The tail is almost a foot long, terminating in a tuft. The eyes large and fierce, and its appearance altogether grim, savage, and formidable.

The Bison is clothed on its forequarters with long shaggy hair, forming a beard beneath its lower jaw, and descending below the knee in a tuft, forming a dense mass on the top of his head, which is so thickly matted as to cause a rifle ball to rebound, or lodge only in the hair by deadening its force. The ponderous head, rendered terrific by this thick shaggy hair, is supported upon a massive neck or shoulders, the apparent strength of which is more imposing from the augmentation produced by the hump, and the long fall of hair by which the fore parts of the body are covered.

It is peculiar to America, formerly inhabiting the prairies and forests in vast numbers; they have been seen in herds of three, four, and five thousand, blackening the plain as far as the eye could reach. They generally seek their food in the morning and evening, retiring during the heat of the day to marshy places. When feeding they are often scattered over a vast surface; but when they move forward in mass, they form a densely impenetrable column, which once fairly in motion can scarcely be impeded, though their unwieldy appearance would indicate slight power of locomotion. They swim large rivers in the same way in which they traverse the plains. When flying from their pursuers, it is impossible for the foremost to halt, since the herd rushing on in the rear, the leaders must advance, though destruction await them. Of this the Indians avail themselves, and no method could be better devised to destroy them, than that of forcing a herd to leap together from the brink of a precipice.

It may not be uninteresting to relate how this is done. One of the swiftest and most active young men is selected, who disguised in a bison skin, having the head, ears, and horns adjusted on his own head, stations himself between the bison herd and some of the precipices that often extend for miles along the rivers. The Indians surround the herd, and at a given signal rush forward and show themselves with yells. The animals alarmed, and seeing no way open to them but in the direction of the disguised Indian, run towards him, and he taking to flight, dashes on to the precipice, where he suddenly secures himself in some previously ascertained crevice. The foremost of the herd arrives at the brink; there is no possibility of retreat, no chance of escape; they may shrink with terror, but the crowd behind, who are terrified by the approach of the hunters, press forward, and are hurled successively into the gulf where death awaits them. One of the tributaries of the Mississippi derives its name of "Slaughter River" from having been continually used for this purpose.

When the ice is breaking up on the rivers in the spring of the year, the dry grass of the surrounding plains is set on fire, and the bisons are tempted to cross the river in search of the young grass that immediately succeeds the burning of the old. In the attempt to cross, the bison is often insulated on a cake of ice that floats down the river. The savages select the most favorable points for attack, and as the bison approaches, the Indians leap with wonderful agility over the frozen ice to attack him. As the animal is necessarily unsteady, and his footing very insecure on the ice, he soon receives his death-wound, and is drawn triumphantly to the shore.

The numbers of this species are surprisingly great, when we consider the immense destruction of them since European weapons have been used against them: they are however fast disappearing before civilization, equally with the Indian himself; and the time is probably not far distant, when both will only be known in the annals of history. They were once extensively diffused over what is now United States territory, but at the present time their range is very different, being confined to the remote unsettled districts of the north and west, being rarely seen east of the Mississippi, or south of the St. Lawrence. West of Lake Winnipeg they are found as far north as 62°; west of the Rocky Mountains seldom farther north than the Columbia River. The greats plains of the Saskatchewan, and the Red River still abound with them, though the herds are less numerous every year.

The first description given of the Bison is by Thomas Morton, A. D. 1637, in a work entitled "New Canaan." He says, that the Indians " have also made great description of herds of well grown beasts that live about the parts of this lake (Ontario,) such as the Christian world until this discovery hath not been made acquainted with. These beasts are of the bigness of a cow, their fleeces very useful, being a kind of wool, and the savages do make garments thereof," &c.

Mackenzie alludes to a white buffalo, during his explorations, said by the Indians to be numerous in Oregon; this probably was the Rocky Mountain sheep, known to them under that name.

The following account of the last buffalo hunt in 1862, is taken from a Red River paper: "From the Pembina Mountain, the usual rendezvous, the hunters set out about the middle of September—105 riders and some 600 carts. Buffaloes were not found in any numbers till they came near the Little Souris, where they killed 500. Here they stopped a week making pemmican,* in full view of a great number of wolves, who were prowling about in large numbers, and with such audacity, that dozens were seen at a time not half a mile from camp. About 400 of these gentry were captured on the trip. Six hundred fine cows were killed, whereupon the bull's meat, with which they had previously loaded themselves, was thrown away. Scratched faces, sprains, contusions of all kinds fell to the lot of numbers of the hunters. He was a bold rider, and had an extra fine horse, who escaped performing a somersault in these wild reckless races over the ground honey-combed with badger and fox holes, and crannies of all sorts and sizes."

The Bison winters amid the timber and grass of northern Texas, New Mexico, and Arkansas, and by the sources of the Red River, and the Cimarone; half famished and miserable it starts with the springing grass, and in April or early in May turns its face northwards in quest of "fresh fields and pastures new." Travelling in countless legions, sufficient to cover whole townships, driven onward with hunger, it crosses successively the Arkansas, the Smoky Hill, and shows a dark front for miles along the south bank of the Platte; and here it is, that meeting the emigrant trains bound for the Pacific coast, collisions ensue, when thousands of them are shot in mere wantonness by hunters already gorged and overladen with buffalo meat. When food, however, is the object,—and the hides are good for nothing in spring and summer,—cows and calves are marked out for

---

* NOTE.—Pemmican—chopped buffalo meat, pounded with corn, and dried in the sun closely pressed together.

destruction, thus increasing the proportion, already far too great, of surviving males, and dooming the race to earlier extinction. But the white man is by no means his only destroyer. The Indian watches for him in every thicket; by every wooded brook-side, the calf that goes down to quench his thirst, is unwittingly slain by an arrow through his loins. The gray wolf lurks in every hollow, and sneaks through every ravine, watching ravenously for some heedless cow—some foolish calf—some wounded or aged bull to straggle to one side, or fall limpingly behind, where a spring from his hiding place, a snap at the victim's ham-strings, will leave nothing to chance but the appearance of some hungry compatriot to claim a dividend of the spoil. But the wolf and the Indian are not wantonly destructive, they kill to eat, and stop when their appetites are glutted. Civilized man alone kills for the mere pleasure of destroying—the pride of having killed. For thousands of years the wolf and the Indian fed and feasted on the buffalo, yet the race multiplied and diffused itself from the Hudson and Delaware, to the Columbia and Sacramento; from the Ottawa and Saskatchewan to the Alabama and the Brazos. But civilized man with his insatiate rapacity, and enginery of firearms has been on his track, and already his range has shrunk to one tenth its former dimensions, and the noble brute is palpably doomed to early extinction.

CHAPTER XV.

THE MUSK-OX—DESCRIPTION AND HAUNTS—PACHYDERMATA—THE PECCARY—WHERE MET WITH—CETACEA OR WHALES—DIFFERENCE BETWEEN EUROPEAN AND AMERICAN ANIMALS—DIVERSITY OF CLIMATE DEVELOPING SPECIES—PROBABLE EXTINCTION OF CERTAIN CLASSES—CONCLUSION.

The Musk-Ox (*Bos Moschatus*) is found exclusively in the Arctic regions, preferring even there the most barren and desolate parts. Nature, who adapts the wants of her progeny to all emergencies, has paid especial attention to the coat of these animals, by covering them with long dense hair, the inner or fine hair corresponding to the *fur* of the bison; the outer covering, is long thick straggling bushy hair, which envelopes the body, hanging nearly to the ground, and thus gives it a very singular appearance from the shortness of its legs. The eye of the Musk-Ox is very prominent, projecting a considerable distance from the frontal bone. Captain Parry thinks the object of this is, to carry it clear of the quantity of hair required to preserve the warmth of the head, when the terrors and rigors of an Arctic winter are encountered by this sturdy animal.

Richardson, another Arctic explorer, says, they never penetrate the woods,—if woods they can be called, where a few stunted junipers and pines only serve to accumulate the drifting snow,—but procure their food in winter, on the steep sides of the hills which are laid bare by the winds, feeding on the moss and lichens with which the rocks are covered, in lieu of herbage; up these hills they climb with an agility that their massive aspect would not lead one to suppose them capable of. The same traveller says they have no tails; he must however either never have procured this animal, or else taken a very cursory glance of them when procured, for the tail of the Musk-Ox about equals in length that of the bear; though bending inwards and downwards, it is entirely hidden by the long hair on the hind quarters.

The Musk-Ox corresponds in a great measure with the bones of the fossil elk of Iceland, in having rudimentary toes. Its flesh is excellent food; though at particular seasons of the year, the bulls, and the old ones generally, emit a musky odor, which communicates itself to their flesh, rendering it unpalatable; even a knife used in cutting up their flesh becomes so strongly scented, as to require much washing and scouring before it is cleansed; but the females and calves have afforded one of the chief means of sustenance to many of our Arctic exploring parties, when other provisions had failed or given out. It is probable that the muskiness peculiar to some animals arises from some property of their food, or part of their food, which is drawn into their system, as it is an ascertained fact that the root of the *calamus* or sweet scented flag is the exciting cause of the odor of the musk-rat.

From the shortness of their limbs, and the apparent weight of their body, it might be inferred that the Musk-Ox could not run with any speed, but it is stated by Parry that although they run in a hobbling sort of canter that makes them appear as if every now and then about to fall, yet the slowest of them can far outstrip a man. When disturbed they frequently tear up the ground with their horns, and turn round to look at their pursuer, but never commence an attack.

Their horns are employed by the Indians and Esquimaux for various purposes; especially for cups and spoons. From the long hair growing on the neck and chest the Esquimaux make their musquito wings, to defend themselves from those troublesome insects. During August and September they extend their migrations to the North Georgian, and other islands bordering on the northern shores of this continent. By Franklin they were never seen lower than 66° North, but Richardson mentions having seen them as low as 60°.

Pachydermata, or thick-skinned animals, have only one representative on this continent.

Description.—This order includes all non-ruminant hoofed animals, or such as have hoofs whether divided or not, but do not chew the cud.

The only animal of this species indigenous to the country, which finds its way into the Northern Continent, is the COLLARED PECCARY, sometimes called INDIAN HOG (*Dicotyles torquatus*); it derives its name "collared" from a peculiar arrangement of whitish bristles rising up from its fore legs and meeting over its neck, which it has the power of erecting when excited, frightened, or irritated. It closely resembles the common hog in shape, structure, habits, and properties, though not quite equalling it in size. This animal has a great aversion to snakes, and will hunt them out with great avidity; when it sees one of these reptiles it raises its bristles with a most ferocious air, its eyes seem to flash fire, and gathering all its strength, with a succession of quick leaps it brings itself down upon the snake's neck with all four feet together with amazing rapidity, until its victim is exhausted, when it ravenously devours it. These animals are generally met with in herds; and if the hunter ventures to attack or wound one of them when its companions are near, he stands a very good chance of being torn in pieces unless he takes refuge in some tree; and even then, they have been known to surround it, keeping him a close prisoner until succor arrive. They are only met with as we approach the southern latitudes, westward of the Mississippi River; eastward of it they are not known to exist: it is met with in Texas, extending to the Pacific, where the line of its range runs as high as the 33rd parallel, following the isothermal line, thus proving that it cannot endure the rigors of a severe winter. This animal must not be confounded with the wild hog, which like the wild oxen and horses, are the offspring of the cattle left to run wild by the Spaniards nearly two centuries ago, and which have quite assumed the original nature of their species, in the solitudes and canebrakes of the sparsely inhabited districts of the sunny South.

The order *Cetacea* (Whales) though rightly succeeding here, is reserved for another series; for though *Mammalia* in the true sense of the word,—not respiring water like true fishes, but rising to the surface to breathe the atmosphere,—and being, withal, warm-blooded animals, yet their habits and form are so pisciverous, that the series comprising fish and reptiles will commence with this order—it being, as it were, the connecting link between them and the Mammals.

Having enumerated now the different species, we must be struck to find that each animal, with the exception of those peculiar to the country, differs from its European congener, though in some instances so slightly as scarcely to be noticed. There is a singular coincidence between the elevation of temperature and the degree of zoological perfection. The genera of latitudes are often representatives, but never identical. Science is developing the various branches of Natural History more fully than ever it before attempted, and each day, light is thrown upon some peculiarity which has hitherto escaped observation. So far as regards the different titles of the same animal, we leave to nomenclators their disputations about what DeKay has happily termed, "the barren honors of a synonyme;" who, if they make no addition to our already gathered information, at least multiply stories, and republish their own names. The letters and anecdotes collected by the gallant explorers of the remotest districts, and the dwellers on the outmost frontiers of the Far West, are rapidly and surely adding to our knowledge of those parts; for, to their credit, and to the honor of the West Point Military Institution, nine-tenths of all the *correct* information we possess of the geography, geology, topography, and natural history of the farther territories and districts, apart from mere fable, comes from its members and its graduates. From the Hudson Bay territory, where we ought to look for the most information, from their commerce in peltries and contact with rude Nature, very little has been gained; but if the time ever arrives that a railway connecting the waters of the Atlantic and Pacific shall be carried through British America, not only the whereabouts of certain species, now only known as rare in their present haunts, will be definitely ascertained, but facilities will be afforded the scientific enquirer and the sportsman to meet them on their own ground.

The diversity of climate met with in so large a continent naturally adds to the developement of its species; and, according to the different localities traversed, so Nature has, with infinite skill, adapted their inhabitants. For instance, in traversing the mighty prairies the mind assumes the idea of isolation: westward, onwards, without a mountain, with scarcely a hill, with rarely a brook or stream to break the monotony of the barren, dewless landscape; loneliness seems the traveller's concomitant. But a nearer gaze reveals the life there; at intervals the fleet antelope looks shyly down from some distant crest, then is off on the wings of the wind; the gray wolf more rarely surveys him deliberately, and slinks away; the prairie wolf lingers near, safe in his own worthlessness and the traveller's contempt; the funny frisky little prairie dog barks with amusing alarm at his approach, then drops into his hole; slowly, on easy pinion, the hawk circles in the air, and swoops down on the prairie squirrel or the mole; lazily the crow leaps from carcass to carcass, too plethoric to caw, too indolent to be frightened.

Away in the wilds of the Rocky Mountain range, the grizzly bear roams amid his fastnesses, his color harmonizing with the brown, gray and weather beaten crags of his mountain home; while his relatives in the far off icebergs and glacier-clad hills of the Arctic zone is as spotless in his snowy robes, as the prospect around him. Away in the great pine forests of Canada, the black bear and the puma, the lynx and the wolf hold in check the smaller animals which otherwise would overrun the land, whilst they in their turn by feeding on the herbage in its luxuriance, maintain the balance of vegetation, and by their oft forgotten hoards of winter provender, provide a store of nut and fruit bearing trees, to be in their turn the providers for some future race of rodents. Remaining still around the haunts of men are the smaller class always associated with him; and the dread of him acquired at his first approach, becomes to them instinctive and hereditary.

In conclusion, man's necessities and pleasures, as they have been in the eastern hemisphere, will so be here, the cause of great changes: the inaccessible cliffs of the Indian territory will afford a refuge for the goat and the cimaron, but the time is fast approaching when the buffalo, the puma, the beaver, &c., will be known only by ancient records. Man himself, the lord of creation, though he extirpates the noble creatures of the earth, will ever be the slave of the cankerworm and the fly till the time come for the final "restoration of all things," and in the words of the poet, "this mortal shall put on immortality."

CHAPTER XVI.

GLOSSARY OF TERMS.

Although the Author has endeavoured as much as possible to dispense with the difficult terms usually met with in Natural History, there are still some peculiar words which must be introduced for the benefit of beginners; they and their meanings are introduced in the following glossary:

ABORIGINAL.—In its native state,—called by the natives or aborigines.

ACUTE.—Of the sight, or hearing, keen. Of smell, quick. Of the ears or nose, pointed.

ANOMALY.—Irregularity, something out of rule.

ANTERIOR.—Going before, belonging to the fore quarter or fore parts.

CANINE.—Of the dog species, shaped like dog's teeth.

CARTILAGE.—A gristly tough substance.

CRITERION.—A standard whereby anything is judged of.

CYLINDRICAL.—Having a long, round body.

DILATED.—Of the skin, expanded, or widened. Of the eye, with the pupil widely and openly developed.

DORMITORY.—A sleeping place.

EXCRESCENCE.—Some substance growing out of, or on another.

FAUNA.—The animal kingdom.

GASTRIC.—Belonging to the stomach.

HABITAT.—The place where an animal dwells,—its haunts.

HYBERNATE.—To pass the winter.

HYPOTHESIS.—A supposition.

IDENTITY.—Sameness in form and appearance.

INCISOR.—Cutting—a term applied to the teeth.

Indigenous.—Natural to the soil or country.
Integument.—A covering, applied generally to skin.
Isothermal.—Of equal heat or temperature.
Lateral.—Growing out of or running along the sides.
Mammiferous.—That suckle their young.
Molar.—Grinding.
Muzzle.—The snout or nose.
Naked.—Destitute of hair, applied generally to the ears or tail.
Oblique.—Not perpendicular, sloping.
Palmate.—Extended out like the flat of the hand.
Parachute.—Any means of buoying up a falling object, as an open umbrella.
Peltries.—Furs and skins.
Pendulous.—Hanging, not supported below.
Phosphorescence.—Luminosity,—emitting the appearance of light.
Pisciverous.—Feeding on fish.
Posterior.—The hind quarters.
Prehensile.—Capable of grasping with.
Premature.—Before its time.
Protrusion.—Something thrust forward or standing prominent.
Rufous.—Of a reddish colour.
Savannah.—An open meadow in the South.
Scepticism.—Open to doubt and disbelief.
Spiral.—Whirling like a screw.
Transversely.—Running crosswise.
Vermiform.—Shaped like a worm.

INDEX.

	PAGE		PAGE
American Rabbit	78	Dog Family	34
Antelope, Prong-horn	94	Dog, Esquimaux	35
Arctic Fox	38	Dog, Prairie	70
Argali	96	Elk	82
Badger	26	Ermine	30
Bat Family	10	Fisher	29
Bat, Species of	13	Flying Squirrel	75
Bay Lynx	51	Fox, Arctic	38
Beaver	61	Fox, Red	38
Bear, Black	17	Fox, Gray	38
Bear, Grizzly	19	Fox, Black	38
Bear, Polar	22	Glossary of terms	109
Bison	98	Glutton	27
Black-tailed Deer	94	Goat, Rocky Mountain	95
Buffalo	98	Ground Hog	69
Campagnol	66	Hare, The Varying	78
Caribou	88	Indian Hog	105
Carnivora	17	Insectivora	13
Cat Family	46	Jaguar	48
Cat, Wild	51	Jumping Mouse	68
Catamount	48	Lemming	67
Cayota	43	Limitation of Species	7
Cetacea	106	Lynx	50
Cheiroptera	10	Marmot, Quebec	70
Chipmunk	75	Marmot, Franklin's	70
Cimaron	96	Marmot, Hood's	70
Cougar	48	Marmot, Parry's	70
Deer Family	82	Marsupialia	59
Deer, Red	91	Marten	29
Deer, Rein	88	Meadow Mouse	66
Deer, Common	91	Mink	30
Deer, Black-tailed	94	Mole Shrew	13

INDEX.

	PAGE		PAGE
Mole, Star-nosed	15	Ruminantia	80
Moose	86	Sable	29
Mouse	68	Seals	53
Morse	58	Seals, Common	56
Musk-Ox	103	Seals, Hooded	56
Musk-Rat	65	Seals, Harp	57
Musquash	65	Seals, Ursine	57
Mustelidœ	27	Seals, Fœtid	57
Opossum	60	Seals, Great	57
Otter	31	Sheep, Rocky Mountain	96
Pachydermata	105	Shrew Mole	13
Panther	48	Shrew Mouse	16
Peccary	105	Skunk	27
Phocœ	53	Squirrel, Flying	75
Polar Bear	22	Squirrel, Gray	73
Porcupine	76	Squirrel, Black	74
Pouched Rat	68	Squirrel, Fox	74
Prairie Dog	70	Squirrel, Red	74
Prairie Wolf	43	Squirrel, Striped	75
Puma	48	Squirrel, Ground	75
Rabbit	78	Star-nosed Mole	15
Raccoon	25	Stag	82
Rat-tribe	67	Walrus	58
Rat, Common	67	Wapiti	82
Rat, Musk	65	Weasel	30
Rat, Wood	67	Wild Cat	51
Rat, Cotton	67	Wolf, Common	40
Reindeer	88	Wolf, Prairie	43
Red Deer	91	Wolf, variety of	45
Rodentia	61	Wolverine	27
Rocky Mountain Goat	95	Woodchuck	69
Rocky Mountain Sheep	96	Woodrat	67

FINIS.

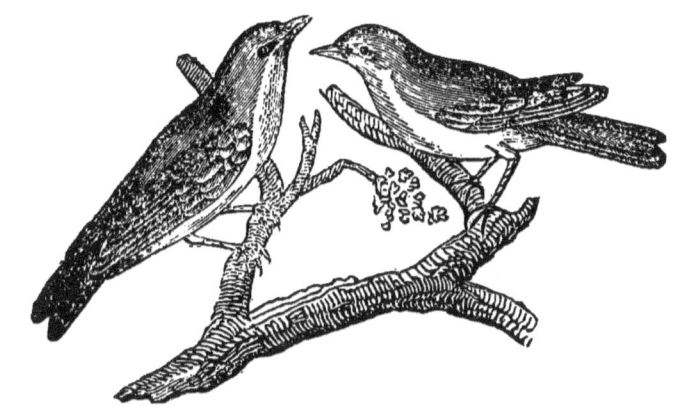

CHARLES A. CRAIG,

Taxidermist,

75 ST. URBAIN STREET, 75

MONTREAL,

ESTABLISHED 1849.

Orders from the Country punctually attended to.

A LARGE ASSORTMENT OF

Rare Canadian Birds and Animals

ON HAND.

ARTIFICIAL EYES, GLASS CASES, &c.

GUILBAULT'S
Botanic and Zoological Gardens,

OPEN EVERY DAY.

Entrance, Upper St. Lawrence Street, or Upper St. Urbain Street, near the Nunnery.

THIS Establishment contains a delightful Promenade and Lawn, Gymnasium (one of the most complete,) Tight Rope, Quoit Ground, Aerial Swing, DANCING HALL, and a variety of other amusements.

THE MUSEUM

Contains a very large Collection of Curiosities, &c.

The Menagerie
Is one of the largest collections in America of LIVING

WILD ANIMALS, RARE BIRDS & FREAKS OF NATURE.

ALSO,

A GIGANTIC BUILDING,

200 feet by 60 feet, for Skating in winter, and Circus, Concert, Pic-Nic, Ball and Gymnasium, &c., &c. in Summer.

A large assortment of Fruit and Forest Trees, Ornamental Plants, Dahlias, Roses, Poultry, Birds, Animals, &c., always on hand, for sale.

J. E. GUILBAULT, *Manager*.

W. DALTON,

BOOKSELLER, STATIONER,

AND

NEWS DEALER,

CORNER OF CRAIG & ST. LAWRENCE STREETS.

Newspapers, Periodicals,

FASHION BOOKS, NOVELS,

SCHOOL BOOKS,

POSTAGE STAMPS, BILL STAMPS,

AND EVERY THING IN THE

PERIODICAL AND STATIONERY

TRADE,

FOR SALE BY

W. DALTON,

Corner Craig and St. Lawrence Streets,

MONTREAL.

TO SPORTSMEN AND OTHERS.

SYRUP OF BUCKTHORN,

THE OLD ENGLISH PHYSIC FOR

Pointers, Setters, Retrievers, &c.

ALTERATIVE CONDITION POWDERS,
GARGLING OIL,

Blistering  and Purging
Ointment, Balls,

FOR

HORSES, CATTLE & SHEEP.

Recipes for HORSE MEDICINES carefully prepared. Orders from the Country attended to with despatch.

HENRY R. GRAY,

DISPENSING CHEMIST,

94 ST. LAWRENCE MAIN STREET, MONTREAL,

(Established 1859.)

www.ingramcontent.com/pod-product-compliance
Lightning Source LLC
Chambersburg PA
CBHW020139170426
43199CB00010B/817